LIGHT SOURCES

Second Edition

LIGHT SOURCES

Second Edition

Basics of Lighting Technologies and Applications

Spiros Kitsinelis

CRC Press
Taylor & Francis Group
Boca Raton London New York

CRC Press is an imprint of the
Taylor & Francis Group, an **informa** business

CRC Press
Taylor & Francis Group
6000 Broken Sound Parkway NW, Suite 300
Boca Raton, FL 33487-2742

First issued in paperback 2017

© 2015 by Taylor & Francis Group, LLC
CRC Press is an imprint of Taylor & Francis Group, an Informa business

No claim to original U.S. Government works

ISBN-13: 978-1-4822-4367-3 (hbk)
ISBN-13: 978-1-138-03404-4 (pbk)

Visit the Taylor & Francis Web site at
http://www.taylorandfrancis.com

and the CRC Press Web site at
http://www.crcpress.com

Dedicated to my son Christos…the light of my life!

Contents

Prologue

The spread of electricity and the development of power grids that began with Thomas Edison changed the way we illuminate our lives and created a new industry. The present book, just like its first edition, provides an overview of the three main technologies that gave birth to the numerous families of products one can find in the market today. Electrical *incandescence*, electrical *gas discharges*, and semiconductor *light-emitting diodes* are those three dominant technologies that have lighted our world for more than a century.

This second edition not only provides updates on the scientific and technological developments of existing and new light sources but also expands on the topic of applications. Different lamp technologies are characterized based on a large number of parameters that need to be taken into account before making a choice; but in this book, the selections and proposals for various applications are based mainly on the "quality" of light, which basically means the *color/wavelength* of the light, the *color rendering index* of the source, and in the case of white light— the *color temperature*. In some cases, other parameters such as the luminous flux or the lifetime may play a role.

This book will assist lighting engineers design the most appropriate environments for people with various needs and professions. Health professionals will be able to create the appropriate visual environment for people with different medical conditions, thus improving the quality of life for many groups of citizens. The book will also serve as a guide for authorities who have to choose the correct technology for cost-effective lighting schemes in order to increase the security and aesthetics of communities. Finally, it will help the ordinary citizen decide which technology will suit him or her the best at home or at work.

Other Sources

This book is an overview of all the main technologies and important families of light sources that have dominated the market and our lives since the end of the 19th century. It is not, however, an exhaustive list of all the available commercial products. There is no point in trying to keep up with all the special characteristics of new products, which change rapidly, and this type of update would be better served by the catalogs of all the major companies.

The photographic material in this book comes from three main sources:

- The author's own collection

- The Museum of Electric Lamp Technology (J.D. Hooker, http://www.lamptech.co.uk)

- Wikimedia Commons (http://commons.wikimedia.org)

The reader will find references at the end of the chapters regarding specific research results, but the general information presented throughout this book is based on the following two important sources:

- *Lamps and Lighting*, 4th Edition, by J.R. Coaton and A.M. Marsden, 1996, London: Routledge/Taylor & Francis Group

- *Electric Discharge Lamps*, by John Waymouth, 1971, Cambridge: MIT Press

More information and publications on lighting design and standards can be acquired from a number of professional organizations such as:

The *International Commission on Illumination* (CIE), which is an international authority and standard defining organization on color and lighting. CIE publishes widely used standard metrics such as various CIE color spaces and the color rendering index (http://cie.co.at).

The *Illuminating Engineering Society of North America* (IESNA) publishes lighting guidelines, standards, and handbooks (http://www.ies.org).

The *International Association of Lighting Designers* (IALD) is an organization that focuses on the advancement of lighting design education and the recognition of independent professional lighting designers (http://www.iald.org).

The *Professional Lighting Designers Association* (PLDA) is an organization focused on the promotion of the profession of architectural lighting design (www.pld-c.com).

About the Author

Spiros Kitsinelis, Ph.D., is a researcher whose focus is on the development of novel and energy efficient light sources and on the communication of this science to a broader audience. He earned his master's and Ph.D. degrees in chemistry from the University of Sheffield in England for his research and development of pulse-operated low-pressure plasma light sources in the High Temperature Science Laboratories. He continued his research as a postdoctoral fellow at Ehime University in Japan in the Department of Electrical and Electronic Engineering. Dr. Kitsinelis held the position of project leader at Philips Lighting Central Development Lighting in the Netherlands, and continued his research and development of the next generation of plasma light sources for the Physics

Department of the National Technical University of Athens, Greece. After a respite from research when he served as a chemical engineer for the armed forces, he acted as the National Contact Point for Energy at the National Documentation Centre of the National Research Foundation of Greece; set up the electronic periodical *Science and Technology of Light Sources (SATeLightS)*; and later worked as a researcher at Paul Sabatier University in Toulouse, France.

Dr. Kitsinelis is the author of a number of scientific publications, has attended many international conferences, and is the cocreator of a number of patents in the field of light sources. His science communication activities include books, television and radio shows, science film festivals, and articles in popular magazines. His latest academic position was that of associate professor at Ehime University in Japan. His personal Web site contains more details at: www.the-nightlab.com.

Dr. Kitsinelis has previously published two books with CRC Press/Taylor & Francis: *Light Sources: Technologies and Applications,* First Edition (2010), and *The Right Light: Matching Technologies to Needs and Applications* (2012).

Acknowledgments

For this second edition I would like to thank my editor Luna Han at CRC Press/ Taylor & Francis for her efforts; my Ehime University colleague, Professor Masafumi Jinno, for his guidance in all professional matters; and finally the readers of the first edition of *Light Sources* who supported the publication.

Introduction

Humanity has to deal with two main issues regarding energy. The first is the availability of nonsustainable energy sources and whether the global demand for energy can be met. This is due to energy source depletion in certain parts of the world or due to geopolitical factors, and in any case, the impact to the global economy is substantial. The second issue is one that deals with the environmental changes of the planet and the impact these changes have on our lives. The burning of fossil fuels as the most common energy generation mechanism results in the formation and emission of carbon dioxide as a byproduct, which is one of the gases responsible for the greenhouse effect.

Considering that humans are using about a fifth of the world's generated electric energy for lighting applications [1,2], it is easy to appreciate the importance of light source technologies both from an economic perspective and from an environmental standpoint. Light sources and lighting not only represent an economic market of billions of dollars but the consumption of energy for lighting is responsible for the generation of millions of tons of CO_2 gas annually.

Furthermore, light is vital and light sources play an indispensable role in daily life. The quality of life, including aspects such as health and urban security related to traffic and crime prevention, depend on light and its quality. Of course, the use of light sources is not limited to general lighting, but also to a range of other applications that require emissions in the ultraviolet and infrared part of the electromagnetic spectrum, such as sterilization, health science, aesthetic medicine, art conservation, food processing, and sterilization of hospitals or water, to name a few.

Efforts to create light at will, as well as to understand its nature, started thousands of years ago with the use of fire. Over time, the burning of wood was replaced by the burning of oil and later, in the 18th century, by the burning of gas. The harnessing of electricity and its use brought about a revolution not only in the way we live our lives but also in the way we light our lives. Electric lighting technologies have been with us since the middle of the 19th century and have been evolving ever since. The first technology, incandescence of a filament, was due to the efforts of people such as Heinrich Gobel in the middle of the 19th century, and Joseph Swan and Thomas Edison a few years later. The second technology, electrical discharge through gas, became widespread in the beginning of the 19th century thanks to Humphry Davy. The third technology, the use of diodes resulting from developments in the semiconductor field, was born much later in the middle of the 20th century, once again revolutionizing the field of lighting.

When Isaac Newton analyzed white light into its constituent colors in the middle of the 17th century, explaining the formation of rainbows, he did not discount the magic of this phenomenon but opened the door to another magical world that had to do with the nature of light. Even though since the time of the ancient Greek philosophers, questions regarding the way the human eye functions and the nature of light continue to tantalize scientists. Today, after centuries of experiments and scientific disputes, certain ideas and theories have been proven and become universally accepted. Some of the basic properties of light will be discussed in Chapter 1.

References

1. European Commission, Directorate-General for Energy and Transport (2007). EU energy and transport in figures. Statistical Pocketbook 2007/2008. http//ec.europa.eu/dgs/energy_transport/figures/pocketbook/doc/ 2007/2007_pocketbook_all_en.pdf.
2. International Energy Agency (2006). *Light's Labour's Lost: Policies for Energy-Efficient Lighting.* http://www.iea.org/publications/freepublications/publication/light2006.pdf.

① Basic Principles of Light and Vision

1.1 Properties of Light

It is imperative to begin by defining several terms used throughout the book that will help the reader understand the properties of lamp technologies available and how light affects an individual. In addition to the introduction of the basic principles of light and vision, the reader may also refer to the Glossary where definitions of the various terms are provided.

As seen in Figure 1.1, visible light is just a small part of the electromagnetic spectrum to which the human eye is sensitive and it includes waves with lengths from around 380 nanometers (10^{-9} m) to about 780 nanometers. On the lower energy side, the spectrum starts with radio waves used to transfer images and sound (like radio and television) and continues to microwaves, used in devices such as radar and ovens. Further down the spectrum are infrared waves, which are perceived as heat. On the higher-energy side of the visible spectrum with shorter wavelengths is ultraviolet radiation. Thus, X-rays are used in medicine for bone outlining; next are gamma rays and finally cosmic rays. Different regions of the electromagnetic spectrum are also presented in Appendix A.

Since the mid-17th century scientists have been divided regarding the nature of light. One side, which included Isaac Newton, believed in the corpuscular property of light and spoke of the effects such as reflection and shadows to support their views. The other side, which was supported by Christian Huygens, believed in the wave properties of light as shown by phenomena such as diffraction. In the early 20th century, the conflict on the nature of light was resolved when scientists and, in particular, Albert Einstein and Louis de Broglie, provided a new picture of quantum physics by showing the duality of matter and energy at the atomic level.

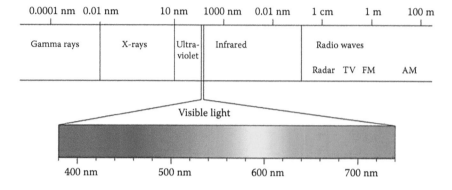

Figure 1.1. Visible light is part of the electromagnetic spectrum and consists of a series of waves with lengths from 380 to 780 nanometers that humans perceive as color.

A series of experiments showed that photons act like particles as demonstrated by the photoelectric effect and Compton scattering but also act as waves. Today, this property continues to be one of the most interesting and bizarre aspects of the natural world.

When discussing the wave property of photons, we talk about electromagnetic waves, which are characterized by intensity, wavelength, and polarization. Electromagnetic waves are transverse waves where the oscillation is perpendicular to the direction of travel, as we know from James Maxwell's equations and the experiments of Heinrich Hertz. Accordingly, an electric field changing over time creates a magnetic field and vice versa. These two fields oscillate perpendicular to each other and perpendicular to the direction of the motion of the wave (Figure 1.2).

Regardless of the wavelength, waves travel at the same speed in a vacuum (299.792458 km/sec) and the electromagnetic spectrum extends from radio waves with wavelengths of up to several kilometers to cosmic rays with wavelengths of

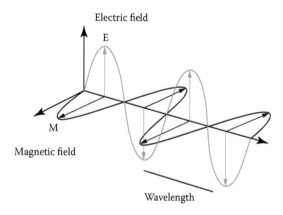

Figure 1.2. Electromagnetic wave propagation.

a few Angstrom (10^{-10} m). The relationship between energy and the wavelength or frequency is given by the formula

$$E = h \cdot v = (h \cdot c)/\lambda$$

where
   E = energy (joule)
   v = frequency (Hz)
   h = Planck constant ($6.626 \times 10{-34}$ J·sec)
   c = speed of light in vacuum (2.998×10^8 m/sec)
   λ = wavelength (m)

Spectrometers analyze light and other radiation by making use of the wave properties of light such as refraction or the interference that comes from diffraction.

The principle of refraction is a change of direction following a change in the speed of the waves that happens when a wave passes from one medium to another with a different optical density, which is called a *different refractive index*. Figure 1.3 depicts this change of direction when the medium changes.

The angle of incidence and the angle of refraction are related to the refractive indices of the media and this relation is described by Snell's law

$$n_1 sin\theta_1 = n_2 sin\theta_2$$

where n is the refractive index of each medium and $sin\theta$ is the sine of each angle (for example, air has a refractive index of 1.0003; for water it is 1.33; and for glass it is 1.5–1.9, depending on the type of glass).

If the angle of incidence exceeds a specific value, then the wave is totally reflected without refraction taking place. This effect, which can be observed only

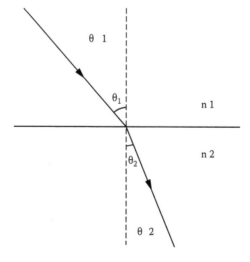

Figure 1.3. Wave refraction.

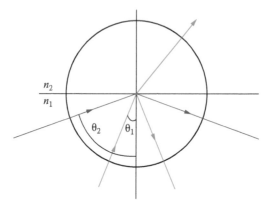

Figure 1.4. Total internal reflection.

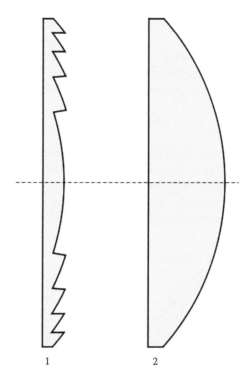

Figure 1.5. Fresnel lens (1) and conventional lens (2).

for waves traveling from a medium of a higher refractive index to a medium of lower refractive index (glass to air, not vice versa), is known as *total internal reflection* and it is the principle upon which optical fibers work (Figure 1.4).

Refraction is behind the optical properties of lenses. The Fresnel design (Figure 1.5) allows for the construction of large lenses by reducing the volume and weight that is required for the construction of conventional lenses. Refraction is

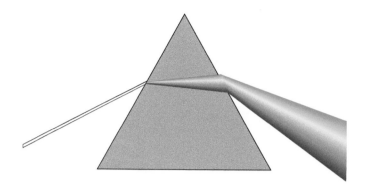

Figure 1.6. Prism diffraction analyzing white light to its constituent colors.

also responsible for the effect of the rainbow and the analysis of white light into its component colors through a prism. The different frequencies of different colors travel at different speeds when they enter a new medium leading to different directions for each wave of a different color. This phenomenon is also called *chromatic aberration* and can be seen in lenses of inferior quality, which means that the wavelength affects the refractive index of the medium. The color of the sky at sunset and changes in the color of the moon are the result of refraction where the Earth's atmosphere acts as the second optical medium. What is important to remember at this point is that—*shorter wavelengths are refracted to higher degrees than longer ones*—so, for example, blue is refracted more than red. Figure 1.6 demonstrates the refraction of white light through a prism. Each color is refracted to different degrees and therefore the waves move in a different direction resulting in the splitting of the different colors.

The other property of light is diffraction. When waves pass obstacles or find an opening of similar dimensions to their wavelengths they spread and interference occurs as shown in Figure 1.7. The interference results in new waves of intensity, which are equal to the sum of the intensities of the initial waves at each point. This relation could mean an increase or the zeroing of intensity at some point of the new wave. Figure 1.8 schematically shows the summation of wave amplitudes.

Figures 1.9 and 1.10 illustrate the effect of diffraction as seen in everyday life and how it is used for the development of scientific instruments.

Another interesting property that will be of value to our discussions is Rayleigh scattering, which is the elastic scattering of light by particles much smaller than the wavelength of the light, such as atoms or molecules. Rayleigh scattering can occur when light travels through transparent solids and liquids, but is most prominently seen in gases. Rayleigh scattering is inversely proportional to the fourth power of wavelength, so that shorter wavelength violet and blue light will scatter more than the longer wavelengths (yellow and, especially, red light). The Tyndall effect, also known as *Tyndall scattering*, is similar to Rayleigh scattering in that the intensity of the scattered light depends on the fourth power of the frequency, so blue light is scattered much more strongly than red light but refers to light scattering by particles in a colloid or particles in a fine suspension [1]. On larger

Figure 1.7. Wave interference.

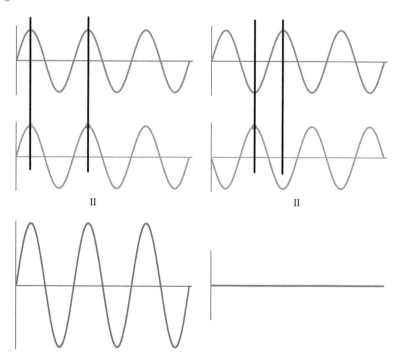

Figure 1.8. The wave interference can be constructive (left) or destructive (right) depending on the phase difference of the waves.

Figure 1.9. When light hits a surface with grooves such as that of a compact disk, diffraction occurs which leads to interference and the formation of colors.

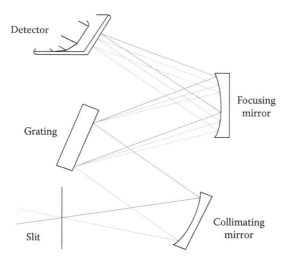

Figure 1.10. Spectrometers use the diffraction effect in order to analyze light into its component wavelengths.

scales (water droplets or large dust particles), there is no wavelength dependence and all light will be scattered or reflected, however, when it comes to outdoor conditions or the eye where scatter can be an issue, then the blue light will be considered more likely to be scattered.

Finally, the kind of light reflection depends on the reflecting material. In specular reflection, the angle of reflection equals the angle of incidence. Materials such as aluminum and silver are used mainly to manufacture mirrors, and show the same reflectivity to all the wavelengths of the visible spectrum. However, for a diffuse (Lambertian) reflection such as the one found with an integrating sphere (Ulbricht spheres), barium sulphate (BaSO4) is used. Integrating spheres are covered in more detail in Section 1.6, "Measuring Light," at the end of this chapter.

1.2   Vision

The way vision works is extremely complicated but, of course, here we will deal with the basic process. The iris of the eye controls the amount of light that passes through and the lens focuses and projects the image upside down at the back end of the eye. A series of sensors are activated and the image is transferred by the use of neurotransmitters to the brain for processing. Problems such as myopia or presbyopia occur when the lens is unable to focus on the right spot or when the size of the eye is such that correct focusing is again not possible (Figure 1.11). The issue of vision will be discussed in greater detail in Chapter 9 (Section 9.4, "Light Sources and Eye Conditions"), when we deal with various eye conditions and lighting aids.

Specifically, the way we see an object is through the following steps: light passes through the cornea, the pupil, the iris, the lens, the vitreous humor, the retina, the optic nerve, the optic pathway, the occipital cortex, and finally reaches the brain where processing takes place. Figure 1.12 depicts the anatomy of the human eye.

The human eye has two different types of photoreceptor cells: cones and rods. Cones are responsible for color perception (there are three different groups of cones for three different parts of the visible spectrum); rods are more sensitive to light intensity, so they are mostly activated at night or in low-level light conditions.

Figure 1.13 demonstrates the sensitivity curves that represent each of the three groups of color-sensing cells in the eyes (the cones). *Cone cells*, or *cones*, are

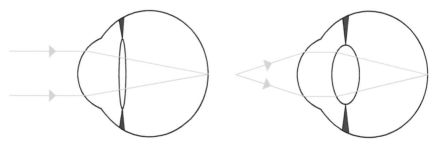

Figure 1.11. Vision problems occur when light from a source is not focused on the correct spot in the eye.

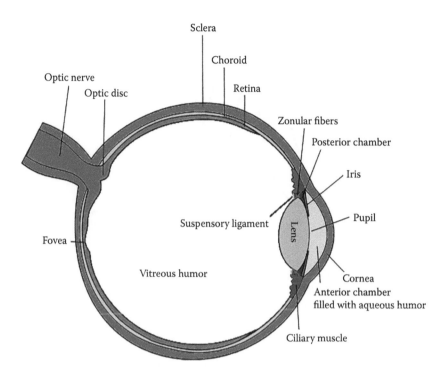

Figure 1.12. Anatomy of the human eye.

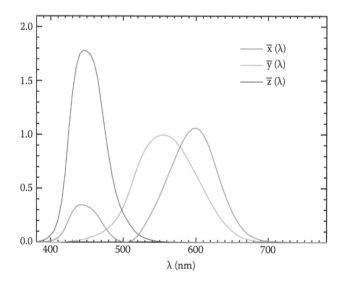

Figure 1.13. Human eye sensitivity curves.

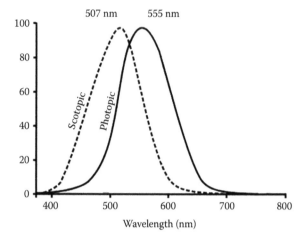

Figure 1.14. This graph shows the sensitivity curve of the eye that peaks at 555 nm. This peak shifts toward the blue when the light is at low levels (rods are more sensitive to blue light). At the maximum sensitivity of the photopic curve, which is at 555 nm, a radiant watt emitted by a source corresponds to 683 lumens (lm).

photoreceptor cells in the retina of the eye that function best in relatively bright light. The cone cells gradually become sparser toward the periphery of the retina. Figure 1.14 illustrates the overall sensitivity of the human eye in different light level conditions.

Thus, the lumen should be considered as a "weighted" power unit taking into account the human eye sensibility. Photopic vision is the only one concerned with lighting design because the eye is "light-adapted" rather than "dark-adapted" under most illumination levels produced by artificial light sources. However, an intermediate situation called *mesopic* vision, which corresponds to dusk conditions, might prove to be a determinant factor for achieving energy savings especially in street lighting systems.

Mesopic vision describes the transition region from rod vision (scotopic) to cone vision (photopic), where signals from both rods and cones contribute to the visual response. Mesopic vision covers approximately four log units and encompasses a range of light levels often found in occupational environments. The performance of light sources is often compared by reference to their efficacy, that is, output measured in lumens per watt of input power. However, the definition of the lumen is derived from the spectral luminous efficiency function for photopic vision only: the eye's sensitivity is described by standards published by the Commission Internationale de l'Eclairage (CIE) for photopic and scotopic conditions. Under photopic conditions, the sensitivity of the human eye peaks at 555 nm. As the luminance decreases, the peak of the sensitivity shifts toward lower wavelengths. The peak sensitivity under scotopic conditions is at 507 nm. These data are known as the *spectral luminous efficiency functions* or the $V(\lambda)$ *curves*. There is not an equivalent standard for the mesopic region, yet developments in this area are forthcoming. The fact that rods are more sensitive to low-light environments, that scotopic vision shifts to shorter wavelengths, and that rod concentration increases in the periphery of the retina mean that rods and peripheral vision are more sensitive to blue light.

Adaptation is the mechanism by which the eye changes its sensitivity to light and can take place by adjusting the iris and pupil size, the nerve endings in the retina, or the chemical compositions of the photopigments in rods and cones. The human eye can function from very dark to very bright levels of light; its sensing capabilities reach across nine orders of magnitude. However, at any given moment in time, the eye can only sense a contrast ratio of one thousand. The eye takes approximately 20–30 minutes to fully adapt from bright sunlight to complete darkness and become 10,000 to 1 million times more sensitive than at full daylight. In this process, the eye's perception of color changes as well. However, it takes approximately 5 minutes for the eye to adapt to bright sunlight from darkness. This is due to the cones obtaining more sensitivity when first entering the dark for the first 5 minutes but then the rods take over after five or more minutes.

Contrast is the difference in luminance between closely spaced areas. Contrast can refer to color; but also, luminance and contrast ratio is the ratio of the higher to lower luminance in the scene. For example, a tunnel that is not very dark may appear very dark when seen from outside and this occurs because the eye cannot be adapted at the same time to very different luminances (see Figures 1.15 and 1.16). When in the field of view something is brighter than what the eye is adapted to then we talk of glare, which can cause discomfort or even an inability to perform tasks. Luminance contrasts must, however, not be too low as the overall result will be a flat visual scene. A well-balanced contrast will achieve

Figure 1.15. The luminance contrast can lead to glare. The photo illustrates this in a tunnel.

Figure 1.16. The luminance contrast can cause areas to appear darker than they really are. The photo illustrates this in a tunnel.

comfort and a satisfactory result. The rule of thumb is that for indoor spaces the contrast ratio should not be higher than 3 and less than 1/3. Lighting requirements for indoor spaces are specified by the European Committee for Standardisation (EN 12464-1), which focuses on maintained luminance levels, uniformity, glare restriction, and color rendering.

1.3   White Light

A wide emission spectrum is called *continuous* while atomic sharp emissions give a line spectrum. Daylight white is composed of all colors, or wavelengths, as shown in the continuous spectrum of Figure 1.17.

Nevertheless, it is possible to create white light without a need for all the wavelengths but with only emissions from each region of the spectrum which stimulate each of the three groups of sensory organs of the eye, namely the cones. More specifically, the impression of white light can be created with a combination of red, green, and blue radiation, these are called the *primary* colors as shown in Figure 1.18. Another way to create white light is by combining blue and yellow light (yellow activates both green and red sensitive cones). This is something that

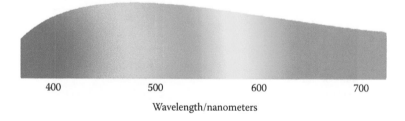

| 400 | 500 | 600 | 700 |

Wavelength/nanometers

Figure 1.17. The color analysis/spectrum of daylight.

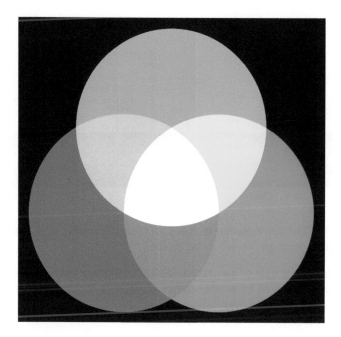

Figure 1.18. The combination of the three primary colors gives white light. Any color can be created with the appropriate combination of primary colors.

light source technologists take advantage of as high efficacy (visible light over total energy consumed by the source) white sources can be developed, although the lack of emissions in many parts of the spectrum means that such sources will not have good color rendering properties.

1.4 Color Rendering Index

The **Color Rendering Index** (**CRI**) is a measure of how well the light source reproduces the colors of any object in comparison to a reference source. The CRI of a lamp is obtained by measuring the fraction of light reflected from each of a number of surfaces of specific color samples covering the visible spectrum. A maximum CRI of 100 is arbitrarily attributed to this light source that most closely reproduces colors as illuminated by standard white light. The less visible parts of the spectrum covered by the radiation source, the smaller the CRI. Some representative values are provided in Figure 1.19 (and Appendix B).

Another issue to be addressed is the color rendering of objects, which is not the same when one uses daylight compared to white light made up from the three primary colors. A color rendering index is therefore defined on a scale of 0 for monochromatic sources to 100 for sources that emit a continuous visible spectrum. The less visible parts of the spectrum covered by the radiation source, the smaller the CRI.

Differences in the color rendering index between two white light sources lead to a phenomena such as *metamerism*. A fabric, for example, whose color is seen

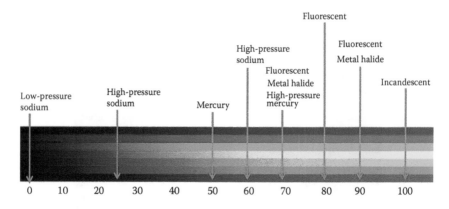

Figure 1.19. The color rendering index increases with increasing proportions of the visible spectrum covered by emissions.

correctly under the light of a source emitting a continuous spectrum, can change color when illuminated by another source of white light that does not emit one of the colors of the fabric dye.

1.5 Color Temperature

How much of the visible spectrum is covered determines the color rendering index, but how the different colors balance is also important. For example, a source that is emitting white light, but with larger percentages of radiation in the red part of the spectrum than the blue gives a feeling of warm white and mostly blue radiation creates a cool white light. This property is called the *color temperature* and is directly linked to the emission of a black body radiation.

A **black body** is defined as a body which has the potential to radiate at all wavelengths with the percentage of radiation being at shorter wavelengths and increasing as the temperature of the body increases. The body initially emits infrared radiation and as the temperature rises, there is emission at wavelengths that the human eye detects. The body initially emits red light, goes through orange, yellow, blue, and finally reaches a white appearance since the emission has covered most of the visible spectrum.

According to Wien

$$\lambda_{max} \ (nm) = 2.8978 \cdot 10^6/T$$

where temperature T is in Kelvin and the wavelength in nm, while the total radiated energy is given according to the Stefan–Boltzmann equation

$$E = \sigma \cdot T^4$$

where E is measured in W/m², T in Kelvin, and σ is the Stefan–Boltzmann constant.

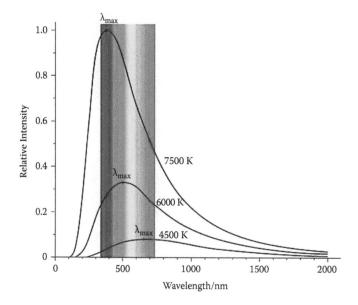

Figure 1.20. The maximum of a black body radiation shifts to smaller wavelengths as the temperature rises.

Thus, the color temperature of a light source is the temperature in Kelvin units in which a heated black body acquires the same color tone to the source. Sources are usually called *warm* when the color temperatures are below 3500 K and cold when at color temperatures above 5000 K (Figure 1.20 and Appendix B).

Different lamps are characterized by different color temperatures and color rendering indices as depicted in Table 1.1 and Figure 1.21 through Figure 1.23.

Companies manufacturing lamps use the terms *warm white*, *cool white*, and sometimes *daylight white*. For the purposes of this book, we define these terms as follows (most companies give similar definitions but not all):

Warm white sources = <3500 K

Neutral white sources = 3500–5000 K

Cool white sources = >5000 K

Table 1.1. Indicative Values of Color Rendering Indices and Color Temperatures for Various Lamps and Light Source Technologies

Light Source	CCT (K)	CRI
Sodium high pressure	2100	25–85
Warm white fluorescent	<3500	55–99
Cool white fluorescent	>5000	55–99
Metal halide	5400	70–90
Incandescent	3200	100

Figure 1.21. Different lamps are characterized by different color temperatures and color rendering indices.

Figure 1.22. Another example of different lamps that are characterized by different color temperatures and color rendering indices.

Figure 1.23. More examples of different lamps characterized by different color temperatures and color rendering indices.

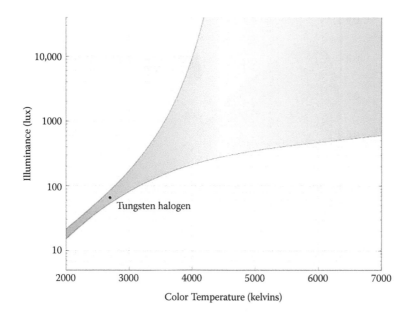

Figure 1.24. The Kruithof curve showing how the color sensation of a given light may vary with absolute luminosity.

Cool white light is sometimes referred to in catalogs as *daylight*.

The color temperatures of these white light sources are defined by the balance between the blue, green, and red emissions in their spectra.

The color sensation of a given light may vary with absolute luminosity, because rods are taking over gradually from cones as the brightness of the scene is reduced. This means, for example, that a cool light source at 6000 K may appear white under high luminance, but appear bluish under low luminance. Under the same low-luminance conditions, the color temperature may need to be adjusted to, say, 4700 K, to appear white. This effect leads to a change in color rendition with absolute illumination levels that can be illustrated in the empirical Kruithof curve [2], which is shown in Figure 1.24.

Since the human eye uses three groups of photoreceptors for three different regions of the visible spectrum, a diagram construction describing all the colors would be three-dimensional. After studies and experiments by W. David Wright and John Guild, the first chromaticity diagram was created. For convenience, the CIE agreed upon two color dimensions (color coordinates) and one dimension of intensity. This way a two-dimensional chromaticity diagram (Figure 1.25) was defined at the maximum intensity of the original diagram. This diagram allows one to see the color that the human eye will perceive from a given light source with a specific emission spectrum.

For some light sources that do not emit continuous radiation throughout the visible part of the spectrum as a black body does, a correlated color temperature (CCT) is assigned, which is defined as the point on the black body locus that is nearest to the chromaticity coordinates of the source (Figure 1.26).

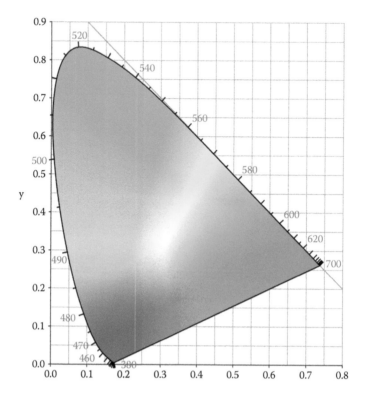

Figure 1.25. The chromaticity diagram.

An incandescent source has a continuous spectrum (with a warm white light due to more red emissions) but there are other ways to produce a white light source. The way to do it is to employ a phosphor that under an excitation emission line from the available source (such as mercury's UV emission lines in a discharge tube or an LED's UV narrow band) will produce visible emissions. The phosphor can be tailored to either produce a wide continuous spectrum if one wants to have white light with a high color rendering index or if one aims for efficiency, then a phosphor is chosen so that it produces only some lines, those that will stimulate the eye's sensors and give the effect of white light (but with low color rendering index).

The way to produce white light with the use of phosphors is discussed both in the discharge light source sections (Chapter 3) and in Chapter 4 covering LEDs. Some of these phosphors are referred to as *tri-* or *tetrachromatic*, which indicates that they are the kind that only produce the three or four narrow emissions required to give a white light effect and high efficacy but have a low CRI. Any conversion of radiation using phosphors includes, of course, losses as higher-energy photons are converted into lower-energy ones (Stokes' losses), and one of the holy grails of light source scientists is to be able to develop phosphors that do not introduce energy differences. This could be achieved if a phosphor was developed that for

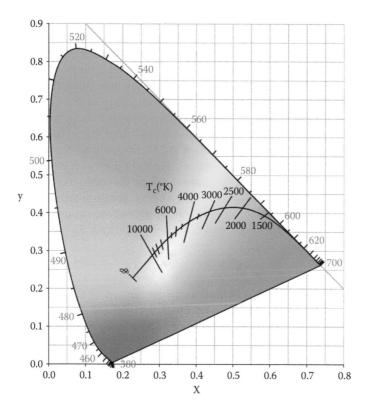

Figure 1.26. The chromaticity diagram with the black body locus and correlated color temperatures for various light sources.

each high-energy photon it absorbs, it would emit more than one photon of lower energy or if a phosphor could be excited by lower-energy photons (red or infrared) giving off a higher-energy photon that would have the energy sum of the stimulating ones. Such a phosphor would revolutionize the light source field as not only would it dramatically increase the efficacy of the sources that already make use of phosphors, but it would also dramatically increase the efficacy of lamps such as incandescent sources where most of the radiation is in the infrared part of the electromagnetic spectrum (where this part is currently unavailable for use in lighting applications). Perhaps this described process will be achieved in the future using different technologies just as quantum dots (see Chapter 4) introduced a new way to convert radiation into different wavelengths.

But using a phosphor that produces the primary colors is not the only way and there is another easier and cheaper way to create white light (as a rule the energy efficient ways to produce white light mean lower color rendering indexes as fewer parts of the visible spectrum are covered by the produced emissions). The three curves that depict the eye sensitivity to the different colors overlap so a yellow emission line (yellow lies between green and red) is able to stimulate both groups of sensors responsible for green and red color. In other words, a source emitting just blue and yellow light will also appear to the human brain as a white source. This source will, of course, have a very low color rendering index as most objects except the blue and yellow ones will not be reproduced correctly. This easy way of producing white light (due to the blue color, the source gives off a cool white light) is exploited in LED technology as the cheapest way to produce white LED sources (a blue LED with a phosphor on top converting some of the blue light into yellow). In the case of discharge tubes, such a technique is also possible (mercury's UV emission at 254 nm could be converted into just blue and yellow) but to the author's knowledge there is no such product, and most phosphors convert UV radiation into more colors. However, this effect has been observed and studied by the author with medium pressure mercury discharges where the visible lines of mercury are significantly enhanced under certain conditions and there is no need for any phosphor (see Chapter 3).

1.6   Measuring Light

The efficiency of a source is defined as the percentage of electrical power converted to radiation and the luminous efficacy is defined as the percentage of power/energy converted to visible radiation. The efficacy is measured in lumens per watt and the theoretical maximum is 683 lm/W if all the electricity is converted into radiation with a wavelength that corresponds to the maximum sensitivity of the eye at 555 nm.

For measuring light from a source, the following terms are useful:

> *Luminous flux*, Φ, defined as the amount of light emitted per second from a light source and the unit is the **lumen** (1 lm = 1 cd·sr) as depicted in Figure 1.27. The measurement of the lumen flux is usually done with an integrating sphere (Figures 1.28 through Figure 1.30), where the light is diffusely reflected by the inner walls.

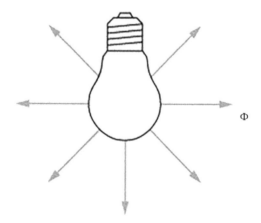

Figure 1.27. Luminous flux Φ.

Figure 1.28. An integrating sphere (Ulbricht sphere) for measuring a lamp's luminous flux.

Figure 1.29. Another integrating sphere (Ulbricht sphere).

Figure 1.30. More examples of integrating spheres (Ulbricht spheres).

Each unit surface of the sphere is illuminated proportionally to the flux of the light source and a small window collects and measures this luminous flux. A unit like the lumen is defined as the total radiated power in watts times the efficacy of the light bulb times 683 lm/W.

The surface of a sphere is $4\pi r^2$ where r is the radius of the sphere, so the solid angle that is defined by the entire sphere is

$$\Omega = 4\pi \text{ (sterad)}$$

The solid angle of a surface A which is part of a notional sphere is defined as

$$\Omega = A/r^2$$

Luminous intensity (I) is defined as the flux of light emitted per second toward one direction and its unit is the candela (cd = lm/sr). It does not depend on the distance. It is measured by a goniophotometer (goniometer), which is shown in Figure 1.31.

$I = \Delta\Phi/\Delta\Omega$, where Φ is the luminous flux and Ω is the solid angle.

One *candela* is defined as the luminous intensity of a source toward a particular direction that emits monochromatic light of a wavelength of 555 nm and a power equal to 1/683 watt per steradian.

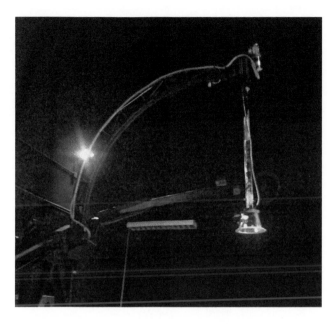

Figure 1.31. A goniometer in a university laboratory.

If a source radiates with equal intensity in all directions, we call it *isotropic* and each time the distance from the source doubles, the intensity measured becomes four times lower; while if the radiation surface is flat, the intensity decreases by the cosine of the angle of measurement according to the law of Lambert (as for luminance), which is schematically demonstrated in Figure 1.32.

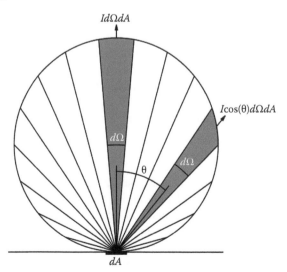

Figure 1.32. Graphical representation of Lambert's law.

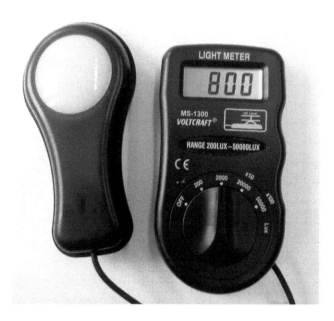

Figure 1.33. A commercial luxmeter.

Illuminance, E, is defined as the amount of light incident on a unit area and the unit of measurement is the **lux** (lx = lm/m²). Illuminance is measured by photometers (Figure 1.33) that convert light into electricity (luxmeter):

$$E = \Phi/A = I/r^2$$

where I is the luminous intensity and A is the area of incidence.

Luminance, L, with the unit cd/m², is defined as the amount of light emitted from a unit surface in a particular direction. It is what is seen by your eyes and it is also defined by the formula L = I/A.

Table 1.2 summarizes the light-measuring terms and definitions.

1.7   Circadian Rhythm

Humans live in a natural environment that has many lighting variations. The day/night cycle, the variations of light during the day, and the changes of the seasons all affect mood and, of course, the activities of all living creatures. In addition, it has been known since ancient times that nonexposure to light and its natural variations is associated with illness and that light can affect moods as well as healing processes. We know that light regulates this internal clock, known as the *circadian rhythm*, by affecting the production of melatonin (relaxation) and cortisol (stimulation) as Figure 1.34 illustrates. The primary circadian "clock"

Table 1.2. The Photometry Units According to the International System (SI)

Quantity	Symbol	Unit	Abbr.	Notes
Luminous energy	Q	Lumen second	lm·s	Also known as *talbots*
Luminous flux	F	Lumen (cd·sr)	lm	Luminous power
Luminous intensity	I	Candela (lm·sr)	cd	SI unit
Luminance	L	Candela per sq. meter	cd/m²	Also known as *nits*
Illuminance	E	Lux (lm/m²)	lx	Used for light incident on a surface
Luminous emittance	M	Lux (lm/m²)	lx	Used for light emitted from a surface
Luminous efficacy	η	Lumen per watt	lm/W	Ratio of luminous flux to radiant flux or total electrical energy

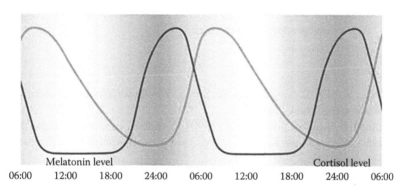

Figure 1.34. Melatonin and cortisol level fluctuations throughout the day.

in mammals is located in the suprachiasmatic nuclei (SCN), a pair of distinct groups of cells located in the hypothalamus. The SCN receives information about illumination through the eyes. The retina of the eye contains photoreceptors (rods and cones), which are used for conventional vision. But the retina also contains specialized ganglion cells, which are directly photosensitive and project directly to the SCN where they help in the entrainment of this master circadian clock [3].

It is principally blue light, around 460 to 480 nm that suppresses melatonin [4], more with increasing light intensity and length of exposure. In fact, research supports that light containing only wavelengths greater than 530 nm does not suppress melatonin in bright-light conditions [5].

The circadian rhythm and the effects of light on it will be discussed in more detail in Chapter 9, "Light Sources and Health."

Having reviewed the properties of light and light sources, the following list covers the principles on which the technology proposals throughout the book will be made:

1. Foveal vision using cones is more sensitive to green and yellow light (photopic eye sensitivity curve peaks at 555 nm).

2. Blue light enhances scotopic and peripheral vision as rods are more sensitive to shorter wavelengths (scotopic eye sensitivity curve peaks at 507 nm).

3. Light of shorter wavelengths (blue light) undergoes more particle scattering.

4. Light of shorter wavelengths (blue light) undergoes greater refraction.

5. A continuous spectrum light source has a higher color rendering index than noncontinuous or near monochromatic sources.

6. A noncontinuous spectrum light source has a lower rendering index but usually higher efficacies than the same power continuous sources.

7. Warm white light creates a cozy and relaxed atmosphere.

8. Cool white light (blue) stimulates us and enhances focus and alertness (suppresses production of melatonin).

References

1. Nave, R. Blue Sky and Raleigh Scattering. http://hyperphysics.phy-astr.gsu.edu/hbase/atmos/blusky.html#c2.
2. Davis, R.G. and Ginthner, D.N. (1990). Correlated color temperature, illuminance level, and the Kruithof curve. *Journal of the Illuminating Engineering Society* 19 (1): 27–38.
3. Brainard, G.C., Hanifin, J.P., Rollag, M.D., Greeson, J., Byrne, B., Glickman, G., Gerner, E., and Sanford, B. (2001). Human melatonin regulation is not mediated by the three cone photopic visual system. *J. Clin. Endocrinol. Metab.* 86: 433–6.
4. Brainard, G.C., Hanifin, J.P., Greeson, J.M., Byrne, B., Glickman, G., Gerner, E., and Rollag, M.D. (2001). Action spectrum for melatonin regulation in humans: Evidence for a novel circadian photoreceptor. *The Journal of Neuroscience* 21 (16): 6405–12.
5. Kayumov, L. (2005). Blocking low-wavelength light prevents nocturnal melatonin suppression with no adverse effect on performance during simulated shift work. *Journal of Clinical Endocrinology & Metabolism* 90 (5): 2755–61.

② Incandescent Lamps

T his chapter is a presentation of the incandescent light sources that have for more than a century been the protagonists in the act of creating artificial lighting. Even today, as discussions are underway related to the banning of this technology due to economic, energy, and environmental reasons, most of these lamps still play an important role. In Europe, this ban will initially include conventional incandescent lamps with high wattage but not reflector lamps and spotlights such as halogen lamps. So, an evaluation of this technology is still very relevant and remains important to the light source professional.

Incandescent light sources have a rich history of developments and discoveries, and have their own unique path of technological evolution and characteristics that make this type of lamp still desirable. At the end of the chapter (Section 2.3, "Overview of the Technology"), some new developments and ideas are discussed that could not only bring this classic technology back in favor but even lead to a new invention and the birth of a new novel light source technology.

The incandescent lamp as the name implies is based on the phenomenon of incandescence. This principle does not differ from that of the black body which radiates as it is being heated, starting from the infrared part of the spectrum and covering more and more of the visible as the temperature increases. Figures 2.1 and 2.2 demonstrate molten glass which begins to radiate in the visible spectrum as the temperature increases. The heating in this case is provided by the electric current flowing through a solid material in the same way the electric oven or an electric kettle operate.

The flow of electricity through a filament increases its temperature as it resists the flow of electrons. The increase in temperature excites the electrons of the

Figure 2.1. An example of visible light emissions from an incandescent material.

Figure 2.2. Another example of visible light emissions from an incandescent material.

filament atoms to higher energy states and as they return to their original state they release this energy by emission of photons.

2.1   Conventional Incandescent Lamps

The name that has been for the most part related to the invention of the electric lamp and more specifically to the electric incandescent lamp is that of Thomas

Figure 2.3. The anatomy of an incandescent lamp.

Alva Edison, while 1879 is referred to as the birth year of this lamp. Thomas Edison was a productive inventor and related his name to the incandescent lamp to such a degree due to some decisive and important contributions that led to the development and commercialization of this product. But the history of this lamp began decades before Edison's time and includes many technologists and scientists from various countries that played important roles. However, the decisive factor in the evolution and spreading of the incandescent lamp was the distribution of electricity to greater parts of the population.

During the first attempts to convert electrical energy into light by means of incandescence in the early 19th century, scientists used platinum and also carbon filaments inside a glass bulb. The bulb was always under a vacuum or contained a noble gas at a low pressure in order to protect the filament from atmospheric oxygen, which accelerated the process of filament evaporation leading to the end of lamp life. After nearly a hundred years of testing, the incandescent lamp took its current form, which employs tungsten filaments. A photograph showing a detailed analysis of the anatomy of a modern conventional incandescent lamp is presented in Figure 2.3.

In brief, some important milestones of the path toward maturity of the incandescent lamp and the people behind them include:

1809—Humphry Davy presents to the Royal Society of Britain the phenomenon of creating light through incandescence by using a platinum filament.

1835—James Bowman Lindsay presents an incandescent lamp in Scotland.

1840—Warren de la Rue heats a platinum filament under vacuum.

1841—Frederick de Moleyns files the first patent for an incandescent lamp with a platinum filament under vacuum.

1845—John Starr files a patent for an incandescent lamp with a carbon filament.

1850 to 1880—Joseph Wilson Swan, Charles Stearn, Edward Shepard, Heinrich Gobel, Henry Woodward, and Mathew Evans develop independently of each other incandescent lamps with various carbon filaments.

1875—Hermann Sprengel invents the mercury vacuum pump. The better vacuum creating technique leads to better lamps.

1878—Thomas Edison files his first patents for incandescent lamps with platinum and later carbon filaments while he also starts distribution of the lamps.

1878—Hiram Maxim and William Sawyer start the second incandescent lamp selling company.

1897—Walter Nernst invents the Nernst lamp.

1898—Carl Auer von Welsbach files his patent for a lamp with an osmium filament. That was the first commercial lamp with a metallic filament.

1903—Willis Whitney develops carbon filaments with metallic coatings thus reducing the blackening of the glass.

1904—Alexander Just and Franjo Hanaman file the patent for an incandescent lamp with a tungsten filament.

1910—William David Coolidge improves the method for producing tungsten filaments.

1913—Irving Langmuir uses noble gases instead of vacuum in incandescent lamps thus making them more efficient and reducing glass blackening.

1924—Marvin Pipkin files his patent for frosted glass.

1930—Imre Brody replaces argon with krypton or xenon.

Incandescent lamps are available in many shapes and sizes while the applied voltage can be from 1.5 to 300 volts. Some of the important advantages of incandescent lamps include their low manufacturing cost and the fact that they can operate with both DC and AC currents. In poor communities, the introduction of other technologies is only possible under strong or total financial coverage by governments or private institutions. Another advantage of this type of lamp is that it functions as a simple resistor so the connection voltage and current are proportional. This means that no additional gear, such as the ballast found with fluorescent lamps, is needed. The lamp accepts the voltage of any household outlet except some special low-power lamps for which the voltage is limited to 12 volts

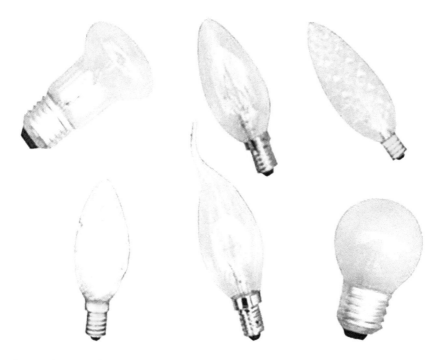

Figure 2.4. Various types of incandescent lamps.

by means of an integrated transformer. Figure 2.4 presents a variety of shapes and forms that incandescent lamps can be found in today while Figures 2.5 and 2.6 show incandescent lamps used with decorative luminaires for indoor lighting.

Attention must, however, be paid to changes in the operating voltage because an increase in voltage will increase the brightness but will also reduce the lifetime and vice versa. Many *long life* lamps are based on exactly this relationship.

The temperature of the filament depends on the voltage while the average life, current intensity, power, luminous flux, and efficiency depend on the filament temperature. Thus, all parameters have a relationship with the voltage, which has been calculated to be as follows:

Current intensity (A) proportional to $V^{0.48-0.62}$

Mean lifetime (hours) proportional to $V^{11.8-14.5}$

Power (watts) proportional to $V^{1.48-1.62}$

Luminous flux (Lumen) proportional to $V^{3.3-4}$

Efficacy (Lumen/watts) proportional to $V^{1.84-1.93}$

The emitted spectrum of these lamps is continuous and covers the entire visible range as illustrated in Figure 2.7. This means that such light sources have a

Figure 2.5. Incandescent lamps are used widely in house lighting luminaires.

Figure 2.6. Another type of an incandescent lamp.

very good color rendering index, which have been designated as 100. The largest percentage of radiation is in the infrared so around 90% of the electric energy is lost in the form of heat. Regarding the 10% of visible emitted radiation, most of it is in red, which gives the white color of the source a warm tone.

Incandescent lamps play a key role in indoor lighting, such as home lighting due to the comfortable warm white light they emit, the variety of shapes they come in, the good color rendering, and their simple operation. However, they are not an economic solution since the efficacy is small (not exceeding 20 lm/W) and the

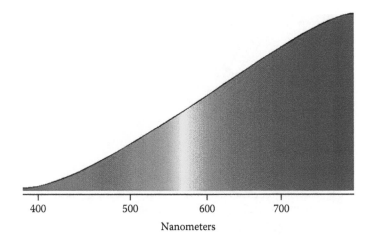

400   500   600   700

Nanometers

Figure 2.7. The emission spectrum of an incandescent lamp is a continuous one.

average lifetime is around 1000 hours. A simple multiplication of the total power of the lamp with the efficacy value gives the luminous flux in lumens. There are also lamps with colored glass bulbs that act as a filter thus giving off light of a specific color or different tones of white. Since the colored glass blocks undesired wavelengths in order to produce the desired color or decreases the amount of red that passes through, the lamp is even less efficient.

The warm light (more red than blue in the emitted continuous spectrum) of operating incandescent lamps used in a typical home are shown in Figures 2.8 through 2.10.

The choice of tungsten was made for the simple reason that it is a metal with a high melting point (3660 K) so it allows the operation of lamps at relatively high filament temperatures (2800 K for conventional incandescent lamp). Higher temperatures mean greater efficacy because a larger percentage of radiation will be in the visible but also a decrease of the average lamp life. Greater power means greater efficacy. For a 5 W lamp with 25 lm flux the efficacy is about 5 lm/W, while for a 250 W lamp with a flux of 4000 lm the efficacy is 20 lm/W.

More support wires make the filament mechanically stronger but reduce the temperature to their thermal conductivity so one has to choose between efficacy and lifetime. Other elements such as thorium, potassium, aluminum, and silicon, or combinations of these, may be added to improve strength. Some lamps have two filaments with three connections at their bases. The filaments share a common ground and can be used together or separately. For these lamps, three numbers are provided—the first two show the consumed power on each filament and the third number is their sum.

To improve the function of the lamp, the filament is in a spiral form, as seen in Figure 2.11, which provides economy of space and heat. The power of the lamp depends on the operating voltage and the resistance of the filament as it is heated and the resistance of the filament depends on the length and diameter. The value of the resistance of a cold filament is about 1/15 of that under operation.

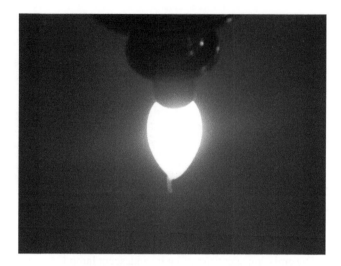

Figure 2.8. An operating incandescent lamp emitting its characteristic warm white light.

Figure 2.9. Another example of an operating incandescent lamp emitting warm white light.

Figure 2.10. An operating incandescent lamp.

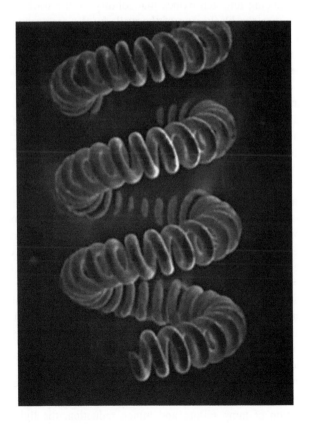

Figure 2.11. A tungsten spiral filament.

For a given lamp power, voltage, and temperature, the filament has a specific diameter and length. Small differences in the resistance value along the length of the filament, which are mainly due to differences in the diameter, result in the creation of hot spots. Tungsten is evaporated faster at these hot spots leading to a dramatic decrease of the lifetime of the lamp. Figures 2.12 and 2.13 show incandescent filaments emitting the characteristic red glow or warm white light.

There are no specifications for the glass of the bulb but the supports of the filament are made of molybdenum and copper so that they melt in case of high currents, thus acting as fuses, and they are enclosed in lead glass that can provide the necessary insulation. The bases of the lamps are usually screw bases, known as Edison bases, having the code letter R followed by a number indicating the diameter in millimeters, or Bayonet bases also having a number indicating the diameter and the code letter B.

In case the bulb is not under a vacuum, the gases contained are argon and nitrogen at low pressures. The main purpose for the inclusion of the gas is to reduce or control the burning of the filament and a more specific reason for using nitrogen is to prevent the formation of sparks.

If there is a leakage in the bulb then the filament comes in contact with atmospheric air producing tungsten oxides that solidify on the walls. If the bulb is under vacuum then the blackening of the walls by the dark tungsten oxides is homogenous. In case the bulb contains a gas such as argon then the oxides are transferred to the top of the bulb due to convection currents. The blackening of the bulb is the second most important factor, after filament vaporization, in reducing the light output. Water is probably the impurity that leads the most to wall blackening and the way it contributes toward that has been named the *water cycle*. Inside the bulb the water molecules break up creating molecules of oxygen and hydrogen. The oxygen molecules react with tungsten and the tungsten oxides that result travel toward the cold spots of the bulb where they deposit. The hydrogen molecules reach those spots and react with the oxides resulting in the tungsten being left on the wall and new water molecules ready to start a new cycle.

The **average rated lifetime** is defined as the time duration beyond which, from an initially large number of lamps under the same construction and under controlled conditions, only 50% still function. Measurements of a rated average lamp lifetime are usually made by applying an operating cycle. For example, the lamps can be operated for 18 hours a day and remain switched off for the other 6 hours or 3 hours on and 1 hour off.

Such measurements offer a good basis for comparisons on technical life and reliability, although the same figures are unlikely to be obtained in practice, because parameters such as supply voltage, operating temperature, absence of vibration, switching cycle, and so on, will always be different.

The **service lamp life** is another term, which is defined as the result of the multiplication of lifetime and lumen maintenance. Often a 70% service life or 80% service lifetime is used. This is the number of operating hours after which, by a combination of lamp failure and lumen reduction, the light level of an installation has dropped to 70% or 80%, compared to the initial value.

Figure 2.12. Incandescent tungsten filaments.

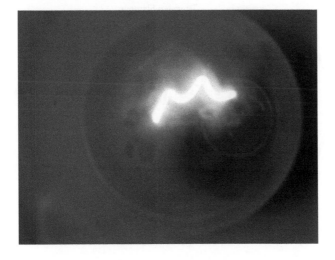

Figure 2.13. Another example of an incandescent tungsten filament.

2.2   Halogen Incandescent Lamps

The halogen lamp is based on the same principle of incandescence but the advantage it offers, compared to the ordinary/conventional incandescent lamp, is that its **average rated lifetime** is twice as long, exceeding 2000 hours. The increase in the average lifetime of the halogen lamp in relation to a conventional incandescent lamp is the result of a chemical balance that takes place within the lamp between the tungsten filament and a halogen gas. The chemical balance is called the *halogen cycle* and the halogen employed is bromine (or compounds of bromine), which replaced iodine that was used in the first halogen lamps. It is because of the inclusion of iodine in the early lamps that the term *iodine lamp* is still used today.

The halogen cycle works as follows: a tungsten atom that has escaped to the gas phase, due to the high temperature of the filament, interacts and bonds with a halogen atom forming a compound that is not deposited at the cold point of the lamp, but continues to travel in the gaseous phase until it again reaches the filament where the high temperature dissociates the compound and the tungsten atom returns to the filament while the halogen atom goes back to the gas phase, ready to start a new cycle (Figure 2.14).

This cycle allows the lamp to operate at higher temperatures (3000–3500 K compared to 2800 K for conventional lamps) with higher luminous efficacy (reaching 30 lm/W), intensity, and color temperature. Because of the higher temperatures required for the halogen cycle to take place, a harder glass such as quartz is used and the dimensions of the lamp are smaller while the filament is thicker. By using a stronger glass, an increase of the gas pressure is also allowed reducing further the vaporization of tungsten. Unlike conventional tungsten-filament lamps, which operate with an internal gas pressure of about one atmosphere,

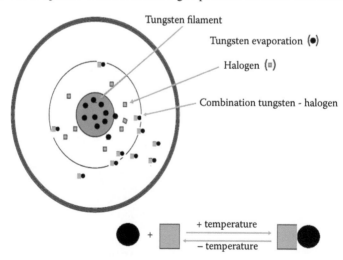

Figure 2.14. The halogen cycle.

most tungsten-halogen lamps operate with an internal gas pressure of several atmospheres to reduce the rate of tungsten evaporation.

Quartz is transparent to UV radiation below 300 nm (the usual soft glass is not transparent to electromagnetic waves with wavelengths shorter than 300 nm) so an additional filter is required if this radiation is not desirable. The additional filter also protects people from glass fragments in case the bulb shatters. Another method of blocking ultraviolet radiation is the use of specially doped quartz glass. A halogen lamp with no filter or special quartz glass can be used as a source of ultraviolet radiation in special scientific applications.

Atmospheres of tungsten-halogen lamps are comprised of an inert gas with about 0.1% to 1.0% of a halogen vapor added. The inert gas may be xenon, krypton, argon, nitrogen, or a mixture of them. The halogen vapor may be pure iodine (I_2) or a compound of iodine (e.g., CH_3I) or bromine (e.g., HBr, CH_3Br, or CH_2Br_2). The minimum bulb wall temperature for the halogen cycle to take place is about 200°C for bromine, which is significantly lower than the 250°C for iodine. Bromine is also colorless while iodine has a very slight absorption in the yellow-green part of the visible spectrum. The amount of halogen that is added is such that it balances the evaporation rate of tungsten at the nominal voltage. An increase in the voltage leads to higher rates of tungsten evaporation rendering the amount of halogen insufficient and leading to wall blackening. If the applied voltage is lower than the nominal one then the lamp temperature might be too low for the halogen cycle to work. At least in this second case, the evaporation rate of tungsten is low enough so that blackening does not occur.

A typical halogen incandescent lamp is shown in Figure 2.15. Attention should be paid to avoid impurities on the surface of the lamp, especially fingerprints that can cause damage to quartz under the high temperature conditions. For the impurities to be removed, the bulb must be cleaned with ethanol and dried before use.

Lamps with reflectors (**R**eflector and **P**arabolic **A**luminized **R**eflectors) can be used as projection lamps for lighting toward a particular direction (Figures 2.16

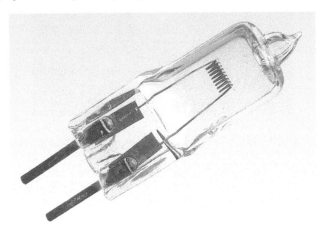

Figure 2.15. A halogen lamp.

Figure 2.16. An example of a halogen lamp with a reflector.

Figure 2.17. Another example of a halogen lamp with a reflector.

and 2.17). In addition to the known bases, halogen lamps have bases with a double contact carrying the code letter G followed by the size of the base in millimeters.

The characteristics of some representative halogen lamp products (Figure 2.18 through Figure 2.21) are tabulated in Table 2.1.

All incandescent lamps produce large amounts of heat, which are emitted together with the visible light toward the desired direction after reflecting from

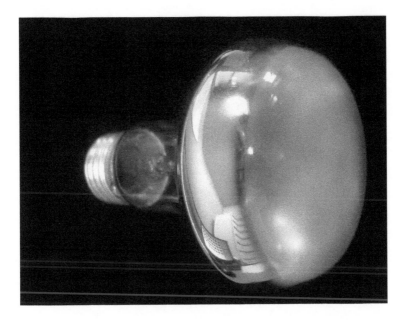

Figure 2.18. There are different geometries for halogen incandescent lamps. This lamp has an Edison screw base, operates with 230 V main voltage, and is omnidirectional.

Figure 2.19. There are different geometries for halogen incandescent lamps. This lamp has an Edison screw base, operates with 230 V main voltage, and is omnidirectional.

Figure 2.20. There are different geometries for halogen incandescent lamps. This lamp has an Edison screw base, operates with 230 V main voltage, and is omnidirectional.

Figure 2.21. This is a spot lamp with a bi-pin (GU) base and operates at 12 V.

Table 2.1. Characteristics of Some of the Lamps on the Market

Figure	Brand	Type	Power/W	CT	Lifetime	Voltage/V	Base
2.18	Osram	Spot R 63					
	42	3000	2000	220–240	E27		
2.19	Philips	Master Classic Eco Boost	30	2930	3000	220–240	E27
2.20	Philips	Ecoclassic	105	2800	2000	220–240	E27
2.21	Osram	Decostar Titan	35	3000	4000	12	GU 5.3

the aluminum reflector of each lamp. If the aluminum reflector is replaced by a **dichroic reflector** then we talk about a cool-beam lamp since at least two-thirds of the infrared radiation passes through the reflector and only the visible radiation and the remaining one-third of the reflected infrared radiation passes through the glass and is emitted toward the desired direction. Therefore, the materials behind a cool-beam lamp, such as the base where the largest proportion of the infrared radiation escapes to, should be able to withstand this exposure to thermal radiation. Of course, the production of heat by incandescent lamps is not desirable in lighting but in other applications where heat is required, these lamps have a significant advantage over other light source technologies. For example, proponents of the incandescent lamp claim that the heat produced balances some of the energy that would otherwise be needed to heat a space in the winter.

Another development, which further increased the luminous efficacy of halogen lamps, is the infrared reflecting coating (IRC). The inside wall of the lamp is covered with several dichroic coatings that allow the transmission of visible light but reflect part of the infrared radiation back to the filament raising the temperature to the desired levels with less energy consumption. The efficiency increases by up to 40% compared to conventional halogen lamps and the average lifetime reaches up to 5000 hours.

Halogen lamps behave in a similar way to conventional incandescent lamps when the applied voltage differs with the specified one. Small increases of the voltage lead to increases of the luminous flux and efficacy but also to decreases of the average lifetime. Each parameter has a different proportionality value but approximately a 10% increase in the voltage value leads to a similar increase of the flux and efficacy and a 50% decrease of the lifetime. In general, for the optimum operation of the lamp, it is best to avoid changes with regard to the specified values by the manufacturer especially when it comes to the applied voltage.

Halogen lamps offer a compromise between conventional incandescent lamps and more efficient compact fluorescent lamps. Halogen lamps still have all the advantages offered by incandescent technology such as instant start and peak brightness, complete control of dimming ratios, absence of harmful materials such as mercury, a lifetime independent from switching frequency, and so forth. On the other hand, they offer additional advantages compared to conventional incandescent lamps, such as a longer lifetime for the same or greater efficacy and a higher color temperature. See Table 2.2 for a comprehensive comparison of various lamp characteristics between different lamp technologies.

Table 2.2. Various Parameter Values for Different Lamp Technologies

	Efficacy lm/W	Power/W	Color Rendering Index	Average Lifetime/Hours
Incandescence	20	15–1000	100	1000
Halogen	30	5–2000	100	2000–5000
Fluorescent	55–120	5–125	55–99	10,000–25,000
Inductive Mercury	70–80	55–165	80	60,000–100,000
Sodium Low Pressure	200	35–180	0	20,000
Xenon Dielectric Barrier Discharge (DBD)	30	20–150	85	100,000
Xenon High Pressure	30	1,000–15,000	90	2000
Sodium High Pressure	50–150	35–1000	25–85	10,000–30,000
High-Pressure Mercury	60	50–1000	15–55	10,000–30,000
Very High-Pressure Mercury	60	100–200	60	10,000
Metal Halide	70–100	35–2000	70–90	10,000–20,000
Sulfur	95	1500	80	60,000 (20,000 driver)
LEDs	100	0.1–10	0–95	50,000–100,000

2.3   Overview of the Technology

Table 2.3 presents some important lamp characteristics and which of the main three light source technologies they offer to a greater degree. Although some important characteristics such as efficacy and lifetime are offered to a greater degree by discharge lamps (fluorescent, neon, sodium, etc.) and solid-state lighting (light-emitting diodes), the incandescent lamps offer advantages in almost all other cases.

Incandescent lamps offer instant start and peak brightness, total control of dimming ratios, absence of any harmful material such as mercury, and the switching

Table 2.3. Comparison of Incandescent Lamp Technology to the Other Two Main Technologies—Plasma and Diode-Based Lamps

	Incandescent Halogen	Electrical Discharges	LEDs
Color Rendering Index	√		
Range of Color Temperatures		√	√
Instant Start	√		√
Lifetime			√
Switching Frequency	√		√
Efficacy		√	√
Cost	√		
Dimming	√		
Operation (AC-DC)	√		
Absence of Extra Gear	√		
Toxic/Harmful Materials	√		
Variety of Shapes—Forms	√		
Range of Power—Voltage	√		

frequency does not affect their lifetime. Other advantages of incandescent lamps are the low manufacturing cost, their ability to operate with both alternating and direct currents, and the fact that extra gear is not required. Their color rendering index is exceptional and this type of lamp technology has many applications, including lighting homes, restaurants, shops, and offices while their variety of shape and form renders them ideal for decorative lighting. Lamps with reflectors (reflectors and parabolic aluminized reflectors) can also be used as spotlights. Although electrical light source technology is the oldest one to penetrate the market, some of its characteristics remain unsurpassed by the other two technologies (discharges and solid-state lighting [SSL]) and offer advantages that enable it to resist newer developments and still play a role.

Research on incandescent lamps continues with most of the work focused on the filament as not only the source of light but also the key element that defines the lifetime of the source. It is expected therefore that novel filaments might lead to a new generation of incandescent lamps with a significantly longer lifetime and efficacy. The research is currently focusing on nanomaterials and specifically the use of carbon nanotubes with the hope that the electrical characteristics of these new materials will offer advantages in a light source. Ideally, the material chosen will be one that manages to convert some of the heat produced into visible light and will also offer mechanical strength and durability. Here are some recently published papers on the subject:

Probing Planck's Law with Incandescent Light Emission from a Single Carbon Nanotube by Y. Fan, S.B. Singer, R. Bergstrom, and B.C. Regan in *Physical Review Letters*, 2009. APS (http://journals.aps.org/prl/abstract/10.1103/PhysRevLett.102.187402).

Brighter Light Sources from Black Metal: Significant Increase in Emission Efficiency of Incandescent Light Sources by A.Y. Vorobyev, V.S. Makin, and C. Guo in *Physical Review Letters*, 2009. APS (http://journals.aps.org/prl/abstract/10.1103/PhysRevLett.102.234301).

Fast High-Temperature Response of Carbon Nanotube Film and Its Application as an Incandescent Display by P. Liu, L. Liu, Y. Wei, K. Liu, Z. Chen, K. Jiang, et al. in *Advanced Materials*, 2009 (http://onlinelibrary.wiley.com/doi/10.1002/adma.200900473/abstract).

Carbon Nanotube Films Show Potential as Very Fast Incandescent Displays (http://www.nanowerk.com/spotlight/spotid = 10776.php).

On another front, research to improve the efficacies of incandescent lamps focuses on *candoluminescence*, the term used to describe the light given off by certain materials which have been heated to incandescence and emit light at shorter wavelengths than would be expected for a typical black body radiator. This phenomenon is noted in certain transition metal and rare earth metal oxide materials (ceramics) such as zinc oxide and cerium oxide or thorium dioxide, where some of the light from incandescence causes fluorescence of the material. The cause may also be due to direct thermal excitation of metal ions in the material. The emission of a bright white light from CaO when heated by an oxygen/hydrogen flame was known as the limelight and its most frequent use was in theaters and during shows to illuminate the stage and performers. The emission of light at energies higher than expected from a flame is also known as *thermoluminescence*. Some mineral substances such as fluorite (CaF2) store energy when exposed to ultraviolet or other ionizing radiation. This energy is released in the form of light when the mineral is heated; the phenomenon is distinct from that of black body radiation.

The most frequently seen case where a substance emits bright white light even under the heating of a flame that does not exceed 2500°C is the case of silicon dioxide (fused silica). It is possible perhaps to fuse together this luminescence phenomenon with the incandescent heating of a simple electric lamp and in a sense fuse together the archaic with the modern way of lighting. The fusion of silica and tungsten could provide the means to generate bright white light even if the laws that govern incandescence and black body radiation only allowed for a yellowish emission of light with low efficiency.

2.4   Codes/Product Categorization

Incandescent lamps are described by various codes that provide information regarding their type, base, and filament as shown in Figure 2.22.

Type of lamp: A letter (or more) indicates the shape of the lamp and a number (or more) shows the maximum diameter of the bulb in eighths of an inch.

Filament type: The letters in front of the code indicates whether the filament is Straight, Coil, or a Coiled Coil while the number indicates the arrangement of the support wires. Figure 2.23 shows various types of incandescent lamp filaments.

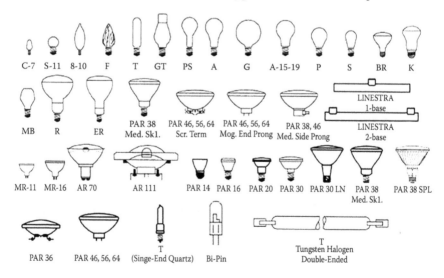

Figure 2.22.   These are the codes of various types of incandescent lamps.

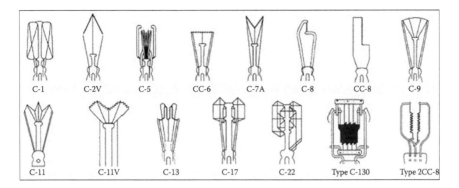

Figure 2.23. Types of filaments.

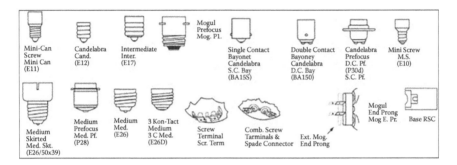

Figure 2.24. Types of bases.

Bases: A letter indicates the kind of base (*E*dison, *B*ayonet, *G*, or *GY* for pins) and a number shows the maximum diameter of the base or the distance between the pins in eighths of an inch. Figure 2.24 shows various types of incandescent lamp bases.

Bibliography

Coaton, J.R. and Marsden, A.M. (1998). *Lamps and Lighting*. London, UK: Arnold Publishing.

Friedel, Robert and Israel, Paul (2010). *Edison's Electric Light: Biography of an Invention*. Revised, Updated Edition. Baltimore, MD: Johns Hopkins University Press.

GE Lighting. Lighting Resources and Information. http://www.gelighting.com/LightingWeb/na/resources/.

Houston, Edwin J. and Kennely, A.E. (1896). *Electric Incandescent Lighting*, New York: The W.J. Johnston Company.

Kaufman, John (ed.) (1981). *IES Lighting Handbook 1981 Reference Volume*, New York: Illuminating Engineering Society of North America.

Klipstein, Donald L. Jr. *The Great Internet Light Bulb Book*, Part I (1996, 2000, 2005, 2006). http://freespace.virgin.net/tom.baldwin/bulbguide.html.

Riseberg, L. Candoluminescent electric light source, U.S. Patent Number 4,539,505 (filed April 29, 1983).

Stark, D. and Chen, A. High efficiency light source utilizing co-generating sources, U.S. Patent Number 6,268,685 (filed August 28, 1997).

Waymouth, John. *Electric Discharge Lamps*. Cambridge: MIT Press, 1971.

③ Electrical Discharge Lamps

3.1 Plasma Processes

This chapter is dedicated to the technology of electrical discharges or plasma for lighting applications. Plasma, which is often called the *fourth state of matter*, is an ionized gas with extremely interesting and useful properties. The term *plasma* was first used to describe an ionized gas by Irving Langmuir in 1927 after observing how the electrified fluid carrying high velocity electrons, ions, and impurities reminded him of the way blood plasma carries red and white corpuscles and germs.

The plasma state here on Earth is quite unusual and exotic but it becomes very common in the outer layers of the atmosphere. It was the development of radio that led to the discovery of the ionosphere, the natural plasma roof above the atmosphere, which bounces back radio waves and sometimes absorbs them. Much of the universe consists of plasma such as our own sun (Figures 3.1 and 3.2), which is a glowing ball of hot hydrogen plasma.

Closer to home, some people have seen a natural plasma in the form of the auroras, a polar light display caused by the bombardment of the atmospheric constituents by high-energy particles, and surely everyone is familiar with the blinding flash of a lightning bolt. It was not long before man appreciated the interesting properties of the ionized gases, and today artificial plasma is being employed in many industrial applications.

One of the most important plasma properties is the emission of light. Photons in a plasma can be produced either by the electron impact excitation of atoms or by recombination of charged particles. This interesting property was identified as early as 1675 by the French astronomer Jean Picard, when he observed a faint glow in a mercury barometer tube. The glow was called the *barometric*

Figure 3.1. Most of the light that shines through the universe originates from plasma sources.

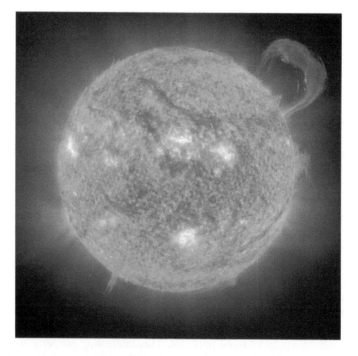

Figure 3.2. Another example of the light that shines through the universe, which originates from plasma sources.

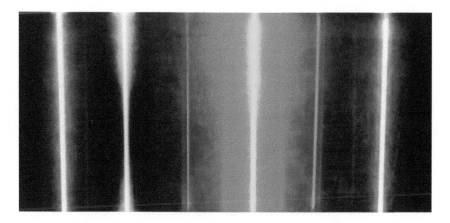

Figure 3.3. Electrical discharges through different gases. Each gas emits radiation at its signature wavelengths. The light emitted from different plasma sources is characteristic of gas or vapor fill.

light and its cause was static electricity. The first development in electric lighting was the arc lamp, which evolved from the carbon-arc lamp demonstrated in 1801 by Sir Humphry Davy, in which an electric current bridges a gap between two carbon rods and forms a bright discharge called an *arc*. By 1855, there was the Geissler tube named after a German glassblower where a low-pressure gas was employed and a voltage was applied across it causing the gas to glow. The neon lamp was developed by the French physicist Georges Claude in 1911, and became very popular due to its red intense emissions. The beginning of the 20th century saw a rapid growth in the development and production of electric discharge lamps but it was not until 1938 that the first practical hot-cathode, low-voltage fluorescent lamp was marketed.

The basic principle is the fact that each atom at the ground energy state is excited, which means that a valence electron leaps to a more energetic state and will return to the ground state by releasing the excess energy in the form of electromagnetic radiation. The photons that are emitted will have energy equal to the one that initially excited the ground atom, that is, equal to the energy state difference (the energy difference between the two orbits that the electron occupied before and after its quantum leap). Such an excitation can take place during collisions with other electrons as in a discharge tube. Each gas emits radiation of different wavelengths [1] as the atomic structure of each element differs (Figure 3.3).

3.1.1 Elastic Collisions

When particles interact in the plasma medium, momentum and energy must be conserved. The collisions that take place can be classified into three different categories. The first type is the elastic collision where momentum is redistributed between particles and the total kinetic energy remains unchanged.

$$e^-_{\text{fast}} + A_{\text{slow}} \rightarrow e^-_{\text{less fast}} + A_{\text{less slow}}$$

It should be noted that the maximum energy a lighter particle can lose through an elastic collision with a heavier atom is 2m/M where m is the mass of the light particle and M is the mass of the heavier one. The efficiency of this energy transfer affects the neutral gas temperature and consequently the cold spot of the discharge.

For example, electrons lose their kinetic energy to atoms more efficiently via collisions with helium atoms rather than xenon. This is one of the reasons argon is used in mercury fluorescent lamps where minimum elastic loss is desired; but neon is used in sodium lamps where higher temperatures are required for the sodium to vaporize.

3.1.2   Inelastic Collisions

The second kind of collision between particles is the inelastic one where momentum is redistributed between particles but a fraction of the kinetic energy is transferred to the internal energy in one or more of the particles. Collisions of this type include electron impact excitation and ionization.

$$e^-_{fast} + A \rightarrow e^-_{slower} + A^*$$

$$\rightarrow e^-_{slower} + A^+ + e^-$$

Light particles such as electrons can lose virtually all their kinetic energy through inelastic collisions with heavier particles.

3.1.3   Super-Elastic Collisions

Finally, a third type of collision can take place where the internal energy of one of the colliding particles is transferred as kinetic energy to the other particles, which results in an increase of the overall kinetic energy of the pair.

$$A^*_{slow} + B_{slow} \rightarrow A_{faster} + B_{faster}$$

The collision cross-sections, which governs the encounter frequency and particle mean-free paths, are energy dependent. For example, high-energy electrons can travel so quickly that the chances of interacting with the outer shell electrons of an atom are reduced.

Some other plasma phase reactions involving charged particles are the following:

$e^- + A^* \rightarrow A^+ + e^- + e^-$	Two-step ionization
$e^- + AB \rightarrow A + B + e^-$	Fragmentation
$\rightarrow A^+ + e^- + B + e^-$	Dissociative ionization
$e^- + A^+ + B \rightarrow A + B + h\nu$	Volume recombination
$e^- + AB^+ \rightarrow A + B + h\nu$	Radiative recombination
$e^- + AB \rightarrow AB^-$	Attachment

$A^+ + B \rightarrow B^+ + A$ Charge exchange

$A^* + B \rightarrow A + B^+ + e^-$ Penning ionization

$A^* + B \rightarrow A + B^*$ Energy transfer

$A^* + A^* \rightarrow A + A^+ + e^-$ Energy pooling

The excited states A^* are metastable states with lifetimes that can last up to milliseconds. Other excited states are not considered since their transition rates are very high (lifetimes of a few nanoseconds).

The parameter, which characterizes a collision process, is the collision cross-section. Electron collision cross-sections depend on impact energy as well as the scattering angles. Experimental cross-sections are difficult to measure so semi-empirical formulas have been developed for different processes such as excitations and ionizations for various plasmas with mercury—argon systems being quite common due to their commercial value.

3.2 Low-Pressure Discharge Lamps

3.2.1 Low-Pressure Mercury Vapor Discharges

The low-pressure mercury vapor discharge has dominated the lighting market for the past few decades and with efficacies reaching 120 lm/W it is only surpassed by the low-pressure sodium discharge (up to 200 lm/W depending on lamp wattage) used mainly in street lighting applications due to its low color rendering index. There are several billions of low-pressure mercury lamps in existence all over the world and the market they represent amounts to billions of dollars annually. The active medium, that is, the element that emits radiation at a low-pressure discharge tube, such as the fluorescent lamp, is mercury, which is a metal in liquid form at room temperature. Mercury, despite its toxicity, is the most commonly used active medium in gas discharges due to its low ionization potential (10.4 eV) and high vapor pressure. Under the operating conditions of the lamp, the pressure of the active medium depends on the temperature of the coldest spot of the lamp, which in this case does not exceed 40°C. At this temperature the pressure of mercury is equal to 10^{-5} of the atmospheric pressure or 7 milliTorr when the atmospheric pressure is 760 Torr (1 Atm = 760 Torr = 760 mmHg = 101300 Pascal = 1.013 Bar).

The passage of electricity through the tube vaporizes mercury and electron collisions with the mercury atoms result in the production of ultraviolet light. This mercury vapor electric lamp, first devised by the American inventor Peter Cooper, but without the phosphor, is the most efficient way of generating light to date. The discharge converts around 60% of the electrical energy into mercury's resonance radiation at the far-UV (253.7 and to a lesser extent 185 nm), which is converted to visible light by a phosphor coated on the inner wall of the glass envelope. It is this efficiency that has made the fluorescent lamp one of the most widely used light sources in the world, creating a huge market. The lamp also contains argon to act as the buffer gas at a pressure of 3^{-5} Torr. The role of the buffer gas is to reduce electrode sputtering, prevent charged species from reaching the walls, and provide an easier breakdown at a lower striking voltage.

Mixtures such as mercury and argon are called *Penning mixtures* due to the proximity of one component's metastable levels (argon's metastable levels at 11.53 eV and 11.72 eV) and the other's ionization potential (mercury's ionization potential at 10.44 eV). Breakdown can occur at lower electric fields for Penning mixtures due to the energy transfer processes that take place. Energy can be stored in the metastable levels of one component and then transferred to an atom of the other component resulting in ionization.

3.2.1.1 The Phosphors

Figure 3.4 defines the anatomy of a low-pressure fluorescent lamp and the steps toward the emission of light. Electrons leave the electrodes, travel through space, collide with mercury atoms, radiation is emitted by the mercury atom, and finally the wall phosphor converts some of that radiation.

Under normal conditions of operation (from mains frequency AC when electromagnetic ballasts are used to a few tens of kilohertz with electronic gear) the most intense emission line of mercury is at 254 nm, which is also known as the *resonance line*, and is the result of an electron transition from the first atomic excited state to the ground state. The process of converting the UV radiation into visible light through a phosphor/fluorescent powder introduces an energy loss mechanism known as *Stokes' losses*, which is not only associated with mercury but with any active medium whose emission photons have to be converted to longer wavelengths with the use of a phosphor powder. In nonluminescent materials, the electronic energy of an atomic excited state is converted completely into vibrational energy so it is lost in the form of heat. In phosphors, the excited state returns to the ground state under emission of radiation, which is almost always at a longer wavelength than the excitation wavelength. This is known as *Stokes' law* and the shorter the excitation wavelength, the larger the energy lost to lattice absorption (Figure 3.5).

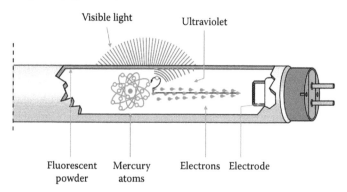

Figure 3.4. Anatomy of a mercury vapor discharge tube known as the *fluorescent lamp.*

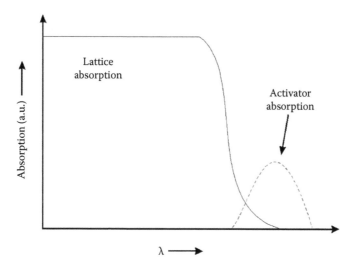

Figure 3.5. The Stokes' losses are minimized if the phosphor excitation source is an emission line that lies close to the visible region. Short-wavelength emissions are absorbed by the lattice and converted into vibrational energy.

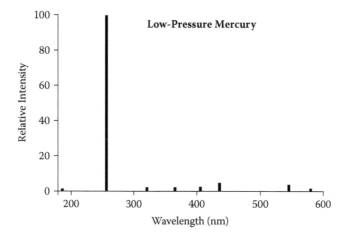

Figure 3.6. The emission spectrum of a low-pressure mercury vapor tube under electrical discharging.

As shown in Figure 3.6, the emission spectrum of a low-pressure mercury vapor discharge tube is not only linear (as opposed to the continuum spectrum), but the largest percentage of the emitted radiation is in the ultraviolet range, namely at 254 nm (see Figure 3.7 for a simplified energy level diagram of mercury).

A large number of fluorescent powders have been developed for many applications and technologies. Appendix C lists a number of these phosphors.

Depending upon the application, the appropriate powder is employed, bearing in mind that improving a characteristic usually compromises another as is

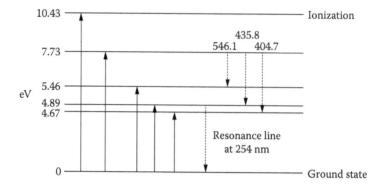

Figure 3.7. Simplified energy diagram of the mercury atom showing the transitions that give rise to mercury's most intense emission lines under a steady-state operation.

the case with the color rendering index and efficiency. For example, if color reproduction is important then a phosphor is chosen that will cover a large part of the visible spectrum such as those listed as broadband/deluxe powders giving the bulb a color rendering index of more than 95. If, however, the luminous efficacy is more important than the color rendering, then one would choose the trichromatic phosphors (these phosphors produce three narrow bands of light in the primary colors). If a warm white color is desired, then a powder is employed that emits a spectrum with emphasis on the red part while for cold white the emphasis shifts to blue.

The first effort to produce visible photons using a phosphor took place in 1934 by G.E. and Osram research laboratories but the major breakthrough occurred with the development of the calcium halophosphate phosphor, $Ca_5(PO_4)_3(Cl,F):Sb^{3+},Mn^{2+}$, in 1942. This type of phosphor combines two complementary color emission bands into one phosphor.

The spectral output of a commercial fluorescent lamp with a phosphor coated on the inside wall can be seen in Figure 3.8. The spectrum illustrates mercury's emission lines and the phosphor's continuum in the red area.

The efficiency of phosphors used today in fluorescent lamps is quite high, exhibiting quantum efficiencies of more than 0.85. The term *quantum efficiency* describes the ratio of visible photos emitted divided by the number of UV quanta absorbed. Nonradiative losses can occur when the emitting state relaxes to the ground state via a crossover of the potential energy curves and they are more likely to happen at higher temperatures.

Of course, there are applications that require UV light such as for sterilizing rooms and equipment. In this case, the resonance line is sufficient so the lamp is used without a phosphor (the bulb in this case should be made of quartz as soft glass blocks radiation with wavelengths below 300 nm due to the sodium carbonate it contains), and tanning lamps which employ powders emitting in the UV-A (310–380 nm).

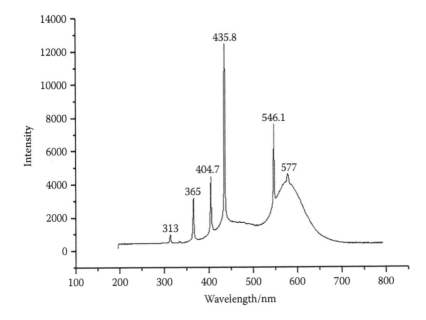

Figure 3.8. Emission spectrum of a low-pressure discharge lamp with phosphor coated on the inside wall.

3.2.1.2 The Electrodes

Electrodes play an important role in starting the discharge. On initially applying a voltage across the tube there is practically no ionization and the gas behaves as an insulator. Once a few ions or electrons are present, a sufficiently high voltage accelerates them to provide more carriers by electron impact ionization and breakdown is achieved. Suitable cathodes supply electrons at a very early stage in the breakdown process by field emission, photoelectric emission, or thermionic emission, greatly reducing the excess voltage needed to strike the discharge. If the electrodes are preheated or if some arrangement, such as an adjacent auxiliary electrode, is provided the voltage required is further reduced.

The electrodes must be good emitters of electrons and must be capable of operating at high temperatures. They should also have a low evaporation rate and high sputtering resistance. This is the reason that tungsten is usually used. But in order to extend the lamp life, the operating temperature should not be very high, so the use of other electron-emissive materials (coated on tungsten) is necessary. The materials most often used are the oxides of calcium, strontium, and barium, which is the coating of the lamp electrodes used in this study (Figures 3.9 and 3.10). The operation of the discharges under AC also ensures a longer life span since both electrodes act as cathodes sharing the workload.

As the triple oxide is very reactive with air, a mixture of the carbonates is coated on the coil, which is inert. The lamps can later be processed under vacuum by heating in order to decompose the carbonates into oxide forms, which have

Figure 3.9. Fluorescent lamp electrodes (tungsten with coating).

Figure 3.10. Another example of fluorescent lamp electrodes (tungsten with coating).

a lower work function and good emissive properties. The maximum gas evolutions for the three carbonate compounds occur at the following temperatures:

$$CaCO^3 = 1062 \text{ K}$$

$$SrCO^3 = 1166 \text{ K}$$

$$BaCO^3 = 1206 \text{ K}$$

Table 3.1 lists the work function values of various compounds used in the construction of discharge lamp electrodes. The work function is the energy needed for an electron to be released from the compound.

Table 3.1. Work Function Values
of Electrode Materials

Cathode Material	Work Function/eV
Tungsten/W	4.54
Barium/Ba	2.5
Barium oxide/BaO	1
Strontium oxide/SrO	1.45
Calcium oxide/CaO	1.75

The dimensions of the electrode and more specifically of the tungsten filament depend on the application of the lamp and its power.

3.2.1.3 The Ballast

An operating gas discharge is a nonlinear circuit element that cannot be described by Ohm's law ($V=IR$) because the conductivity increases with current. The discharge is therefore said to exhibit negative differential voltage-current characteristics, except in some special cases such as weakly ionized plasma. A current limiting device, called a *ballast*, is therefore necessary for the operation of discharge lamps. In a discharge tube, the low-pressure gas between the electrodes acts as an insulator and it is only with the application of high voltage that the gas begins to ionize and becomes conductive. As the current intensity increases, the plasma (ionized gas) resistance decreases due to the larger number of ions so a ballast is needed to control the current to a desired value before it tends to infinity and the tube is destroyed. The function of a ballast is also the application of sufficient voltage to ignite the lamp as well as the regulation of the lamp current, and in the case of AC operation, relighting the lamp at each half-cycle.

A simple series-connected resistor can be sometimes used as a ballast, but the power losses ($I^2 \times R$) result in a low overall efficiency. The use of a resistor on an alternating supply leads to reignition delays causing near-zero current periods at the start of each half-cycle. The advantage offered by AC operation, however, is that an inductive or capacitive impedance can be used to provide current limitation, resulting in a significant energy loss reduction.

Standard fluorescent magnetic ballasts use a combination of inductive and capacitive networks for current control, which essentially comprises an aluminum coil wrapped around an iron core. This combination reduces power losses even more by reducing the phase difference between the voltage and current waveforms. The energy efficient magnetic ballast is an improved version of the standard one employing copper wire instead of aluminum wrapped around larger iron cores. The efficiency is improved due to copper's lower resistance and less heat generated by the larger iron core.

Modern electronic ballasts eliminate the large, heavy iron ballast and replace it with an integrated inverter/switcher. They operate in a similar manner to iron ballasts but at a much higher frequency (>20 kHz instead of 50 Hz that magnetic ballasts operate at). Power is provided to the ballast at the mains frequency and is converted to a few tens of kHz.

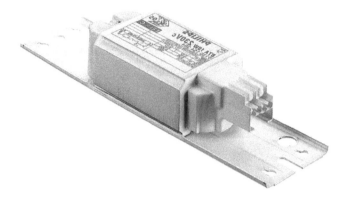

Figure 3.11. Electronic ballast for a fluorescent lamp.

Current limiting is achieved by a very small inductor with high impedance at these high frequencies. The higher frequency results in more efficient transfer of input power to the lamp, less energy dissipation, and elimination of light decay during each cycle. For economic reasons a combination of a standard and an electronic ballast is manufactured, usually referred to as a *hybrid ballast*, offering good energy efficiency but reduced light output and lamp life. The trends in the market show that electronic ballasts, such as that shown in Figure 3.11, will dominate, and soon all discharge lamps of any pressure will employ them.

Along with controlling the intensity of current to a desired value, the ballast/gear also manages the starting of the lamp, which is achieved in two ways. In one case, the electrodes are heated, thereby reducing the initial voltage needed for ignition, but this way the lamp consumes more energy. In the other case, the electrodes are not heated, thus saving energy, but the applied voltage needs to be higher so the electrodes are subject to greater wear thus reducing the average life of the lamp. The choice of starter depends on the use of the lamp, so when multiple switches are involved, the heated electrodes are preferred. Finally, the electronic ballast offers the option of dimming all the way down to a small percentage of the maximum brightness.

The tubular fluorescent lamps, such as those in Figure 3.12, come in various lengths depending on the wattage of the lamp and they are known by a code (such as TL) followed by a number indicating the diameter of the tube in eighths of an inch. The most widespread are the TL8 (one inch diameter) but they are losing popularity to the TL5. Figure 3.13 through Figure 3.22 show representative commercial fluorescent lamps and their typical emission spectra while Table 3.2 through Table 3.6 show some of their typical lamp characteristics. Throughout the whole chapter there will be emission spectra and characteristics tables of representative discharge lamps.

Another broad category of lamps is the compact fluorescent lamps (CFLs), which are increasingly used as replacements for incandescent bulbs due to their energy savings. The placement of the electronic ballast at the base of the lamp makes them even more practical.

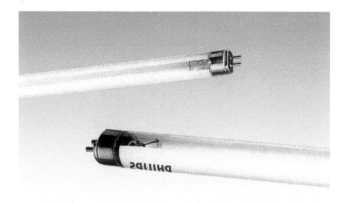

Figure 3.12. Tubular fluorescent lamps.

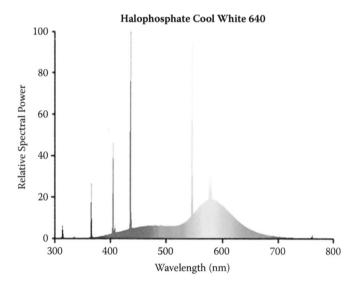

Figure 3.13. Emission spectrum example of a fluorescent lamp with a phosphor.

In general, fluorescent lamps are much more economical than incandescent lamps with a luminous efficacy up to four times higher than the most efficient halogen bulb and a much longer average life. For this reason, they are the preferred choice for applications where color rendering has to be good but not great, high brightness is required, and replacement is not easy (the induction lamp, in particular, is useful for this latter reason). However, it must be noted that there are fluorescent lamps on the market that demonstrate near perfect color rendering properties with the appropriate phosphors but at the expense of efficacy. Fluorescent lamps have dominated the lighting market and are mainly used for general lighting in offices, living spaces, and other large areas such as warehouses.

Figure 3.14. Circular fluorescent lamp (Sylvania circline ring-shaped 60 W).

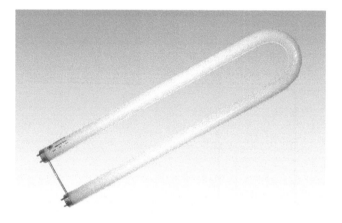

Figure 3.15. U-shaped fluorescent lamp (GE U-Shaped 40 W).

From the fluorescent lamp, which is a low-pressure mercury vapor discharge tube, the next category of lamps operates on the same principle of low-pressure electrical discharges, but the active media are different, thus giving different emission spectra.

3.2.2   Low-Pressure Noble Gas Discharges

If we replace mercury with a range of gases such as the noble gases then we have discharges with instant starting because the active medium is already in gas phase and is not affected by ambient temperature or the temperature of the lamp. Although the noble gases (helium, neon, argon, krypton, and xenon) emit in the visible range, their resonance lines (see Table 3.7), which constitute most of the emitted radiation are in the ultraviolet range and even below 200 nm so appropriate fluorescent powders to convert this radiation into visible light would

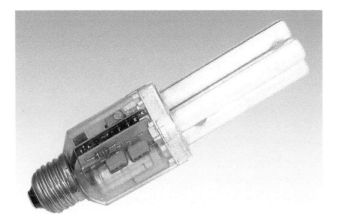

Figure 3.16. Compact fluorescent lamp with integrated electronic gear/ballast (Philips PLCE 7 W).

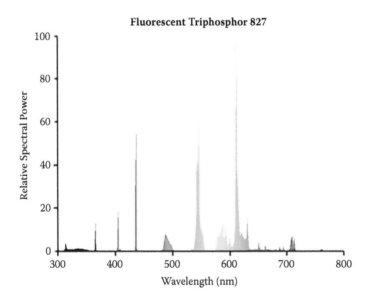

Figure 3.17. Emission spectrum of a compact fluorescent lamp.

be needed. However, the conversion of radiation from UV-C to visible light would mean large Stokes' losses and this is why such lamps are mainly used for decoration and signs and not for general lighting as presented in Figure 3.23.

Of all the noble gases, neon is used most frequently as an active medium in low-pressure discharge tubes. The reason for this is that neon emits a large number of visible emission lines in the red region of the spectrum so neon lamps without employing any phosphors have found applications in advertising signs. Figures 3.24 and 3.25 illustrate neon's red emissions.

Figure 3.18. Compact fluorescent lamp (Philips "Tornado" helical-shaped 23 W).

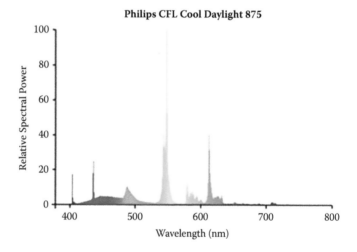

Figure 3.19. Emission spectrum of a compact fluorescent lamp.

Neon is also used in indicator lamps of electronic devices while the rest of the gases are used in more specific applications such as discharge tubes for LASERs. Xenon under pulsed conditions (microsecond scale and frequency of several hundred Hz) and low pressure (several tens of Torr) generates short bursts of high-intensity white light and finds applications in photography (Figure 3.26 and spectrum in Figure 3.27).

A similar technology is that of the dielectric barrier discharge (DBD) that does not utilize the resonance lines of xenon atoms (129 and 147 nm) for the excitation of the fluorescent powder but the emission of xenon dimers (172 nm) generated under the conditions of operation. The visible light emitted by the powder in combination with the visible light emitted by the active medium gives us a light source of low efficacy (30 lm/W) compared to existing fluorescent lamps but of

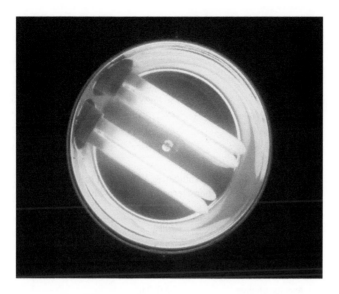

Figure 3.20. Emission of light from compact fluorescent lamps (CFLs).

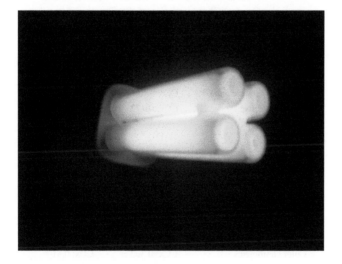

Figure 3.21. An example of a compact fluorescent lamp.

Table 3.2. Indicative Values and Information about the
General Characteristics of the Tubular Fluorescent Lamp

Gas Fill	Ar|Hg	
Luminous flux	5000 lm (100 h)	
Efficacy	100 lm/W (100 h)	
Color temperature and CRI	CCT: wide range	CRI: Ra = >80
Average lifetime	20,000 h	

Table 3.3. Table with Indicative Values and Information about the General Characteristics of the Circular Fluorescent Lamp

Gas fill	Ar|Hg	
Luminous flux	3400 lm (100 h)	
Efficacy	56 lm/W (100 h)	
Color temperature and CRI	CCT: 3000 K	CRI: Ra = 53
Color coordinates	CCx: 0.436	CCy: 0.406
Average lifetime	11,000 h	

Table 3.4. Table with Indicative Values and Information about the General Characteristics of the U-Shaped Fluorescent Lamp

Gas fill	Ar|Hg	
Luminous flux	2875 lm (100 h)	
Efficacy	72 lm/W (100 h)	
Color temperature and CRI	CCT: 3450 K	CRI: Ra = 54
Average lifetime	12,000 h	

Table 3.5. Table with Indicative Values and Information about the General Characteristics of the Compact Fluorescent Lamp

Gas fill	Ar-Kr|Hg	
Luminous flux	400 lm (100 h)	
Efficacy	57 lm/W (100 h)	
Color temperature and CRI	CCT: 2700 K	CRI: Ra = 82
Color coordinates	CCx: 0.473	CCx: 0.420
Average lifetime	8000 h	

Table 3.6. Indicative Values and Information about the General Characteristics of the Compact Fluorescent Lamp

Gas fill	Ar|Hg	
Luminous flux	1450 lm (100 h)	
Efficacy	63 lm/W (100 h)	
Color temperature and CRI	CCT: 7350 K	CRI: Ra = 83
Color coordinates	CCx: 0.303	CCx: 0.310
Average lifetime	6000 h	

long average life (over 100,000 hours) and it offers the advantage of avoiding environmentally harmful materials such as mercury (Table 3.8). This technology (DBD xenon) has already been channeled into the market in the form of a flat light source (surface area of 21 inches) for general lighting or liquid crystal display backlight (Figures 3.28 and 3.29).

Based on the emissions of the xenon dimer molecules under pulsed conditions of operation and the use of a fluorescent powder to convert ultraviolet light into visible light, an electroded discharge tube has been developed for general lighting

Table 3.7. Table with Ionization Energies and Resonance Line Wavelengths of Various Active Media

Element	Metastable Levels/eV	Ionization Energy/eV	Resonance Emission Lines/nm
Neon	16.53, 16.62	21.56	73.6, 74.4
Argon	11.49, 11.66	15.76	104.8, 106.7
Krypton	9.86, 10.51	13.99	116.5, 123.6
Xenon	8.28, 9.4	12.13	129.5, 146.9
Mercury	4.64, 5.44	10.43	184.9, 253.7
Sodium	—	5.14	589.0, 589.6

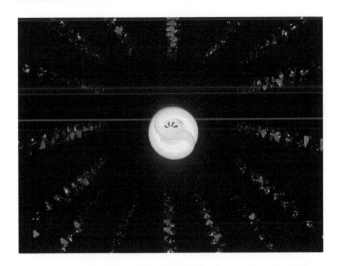

Figure 3.22. A compact fluorescent lamp.

Figure 3.23. Low-pressure gas discharge lamps are used for decoration and signs.

Figure 3.24. The neon discharge has dominated the advertising sign industry for decades.

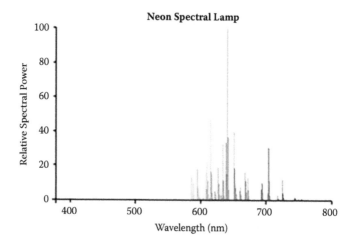

Figure 3.25. Low-pressure neon discharge lamp emission spectrum.

Figure 3.26. Photographic xenon flash during light emission.

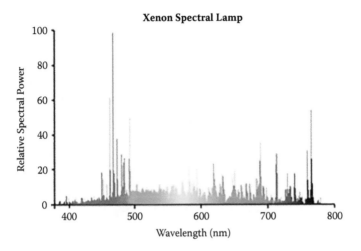

Figure 3.27. Low-pressure xenon emission spectrum.

Table 3.8. Indicative Values and Information about the
General Characteristics of the Low-Pressure Xenon Lamp

Gas fill	Xe_2	
Luminous flux	1850 lm (100 h)	
Efficacy	27 lm/W (100 h)	
Color temperature and CRI	CCT: 7400 K	CRI: Ra = 86
Color coordinates	CCx: 0.304	CCy: 0.295
Average lifetime	100,000 h	

Figure 3.28. Dielectric barrier discharge (DBD) lamp with xenon at low pressure
for general lighting (Osram Planon 68 W).

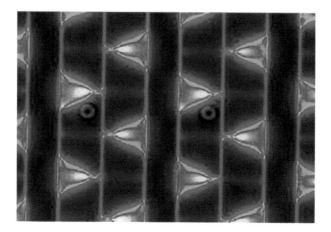

Figure 3.29. Plasma formation in a xenon DBD lamp.

although the efficacy is still low compared to low-pressure mercury vapor discharge lamps. An interesting development of this technology is the use of multiple electrodes by Jinno, Motomura, and Aono [2] in order to maintain plasma diffusion and homogeneity and prevent the plasma inside the tube from constricting.

Dielectric barrier discharge (DBD) excimer or exciplex lamps are known as efficient UV and vacuum ultraviolet (VUV) sources with a narrow spectrum. Another example of DBD development is the Xe-Cl$_2$ DBD lamp producing more than 90% of its radiated power at 308 nm, which corresponds to the emission of the Xe*Cl exciplex molecule.

There is ongoing research and development of xenon discharge lamps under pulsed operation for general lighting due to the pressure to ban environmentally harmful materials such as mercury; and several references in scientific journals reveal that their luminous efficacy seems to be increasing.

3.2.3 Low-Pressure Sodium Vapor Discharges

The low-pressure discharge lamp with the highest luminous efficacy that reaches 200 lm/W is that of sodium vapor. This high efficiency is due to the fact that the resonance line of sodium (Figures 3.30 and 3.31), produced after excitation of sodium atoms by electron collisions, is at 589 nm, that is, it is in the visible spectrum and does not need conversion by a fluorescent powder.

The lamp is made of hard boron glass to withstand the corrosiveness of sodium and also contains a mixture of neon/argon, which acts as the buffer gas. The partial pressure of sodium is much lower than that of mercury so for the proper operation of the lamp higher temperatures are required than that of a fluorescent lamp. To increase the temperature a second outer tube encloses the inner one in a vacuum and a special coating is applied on the inner walls of the outer tube that allows the escape of visible light but reflects back the infrared radiation (heat).

One of the reasons the low-pressure sodium lamp has a high efficacy is due to the fact that sodium's resonance line lies around that peak of the human eye's sensitivity curve (589 nm), making it at the same time inappropriate for domestic

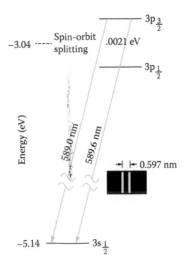

Figure 3.30. Simplified energy level diagram of sodium showing the resonant states.

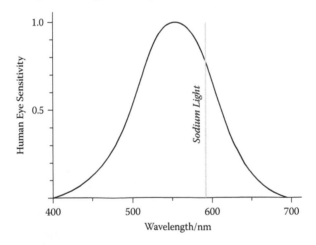

Figure 3.31. Sodium's resonance line lies around the peak of the human eye's sensitivity curve (589 nm).

lighting due to its poor color rendering properties. The color rendering index (CRI) is a relative measure of the shift in surface color of an object when lit by a particular lamp, compared with how the object would appear under a reference light source of similar color temperature.

The low-pressure sodium lamp has a color rendering index of zero (and low color temperature of around 1700 K) so it is mainly used in applications where high brightness is required but color reproduction is not so important, such as street lighting, safety lighting, lighting of large outdoor spaces (parking spaces), and so on. Figures 3.32 and 3.33 provide examples of street lighting employing low-pressure sodium vapor discharge lamps.

Figure 3.32. Example of low-pressure sodium vapor discharge lamps for street and road lighting.

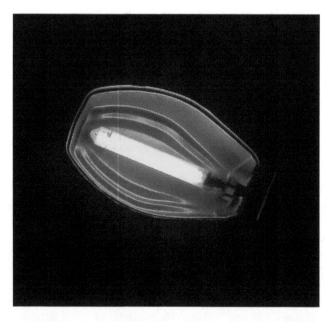

Figure 3.33. Another example of a low-pressure sodium vapor discharge lamp.

A good candidate to replace existing active media in discharge lamps must be an efficient emitter in the range of 380–580 nm. The reason for selecting this particular range is the need to eliminate any Stokes' losses associated with photons that the human eye cannot detect and would need conversion with the use of a phosphor. On the short-wavelength side, the eye sensitivity and therefore the visible range starts at around 380 nm. On the long-wavelength side, perhaps there is no need to include wavelengths beyond 580 nm as sodium is already a strong emitter; just above that limit the sensitivity of the eye drops and, in any case, the photons on that part of the spectrum cannot be converted with the use of phosphor yet. On the other hand, strong emissions in the blue region could be partly converted in order to cover larger regions of the visible light.

The average lifetime of these lamps reaches nearly 20,000 hours. Until the temperature reaches the required levels where the necessary amount of sodium vaporizes, light is emitted by the excited buffer gas neon atoms, which explains the red color of the lamp light during its first few minutes of operation after each start. As sodium vaporizes and more of its atoms participate in the electrical discharge process, the lamp gets its characteristic yellow color due to radiation at 589 nm. Figure 3.34 through Figure 3.36 provide a representative commercial low-pressure sodium vapor lamp along with a spectrum; while Table 3.9 provides general information about the low-pressure sodium vapor lamp.

3.3   Pressure Effect

3.3.1   Thermal Equilibrium

All plasma has a number of features in common but one way to begin classification is by quantifying the density of the charge carriers and the thermal energy of the electrons, in particular. One of the plasma parameters that has a pronounced effect on both of these quantities is the pressure of the gas.

Figure 3.34. Low-pressure sodium vapor discharge lamp (Philips SOX 35 W).

Figure 3.35. Electrode of a low-pressure sodium vapor discharge lamp.

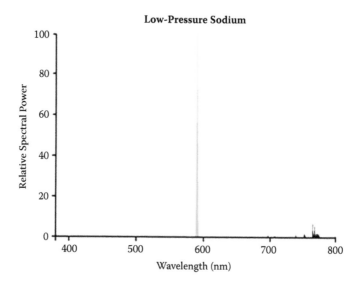

Figure 3.36. Emission spectrum of a low-pressure sodium vapor discharge lamp.

Table 3.9. Table with Indicative Values and Information about the General Characteristics of the Low-Pressure Sodium Vapor Lamp

Gas fill	Inner: 1% Ar in Ne|Na	Outer: Vacuum
Luminous flux	4800 lm (100 h)	4560 lm (end of lifetime)
Efficacy	126 lm/W (100 h)	120 lm/W (end of lifetime)
Color temperature and CRI	CCT: 1700 K	CRI: Ra = 0
Color coordinates	CCx: 0.574	CCy: 0.425
Average lifetime	20,000 h	

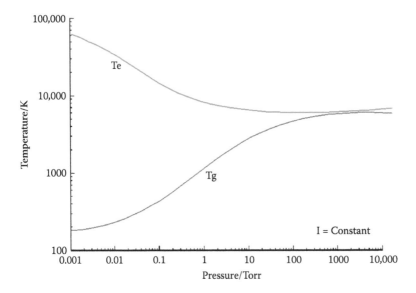

Figure 3.37. Gas and electron temperatures with respect to gas fill pressure. At higher pressure, conditions of local thermal equilibrium are achieved.

One of the principle characteristics of a low-pressure discharge is the absence of thermal equilibrium in the plasma. While a low-pressure discharge is a collision-dominated plasma, the collision rate is generally not high enough for the electrons to equilibrate with the heavy atoms. In all cases, the electron temperature is much greater than the kinetic temperature of the atoms, ions, and molecules. An average value for the electron temperature in low-pressure discharge is 11,000 K, which corresponds to 1.0 eV, but a fraction of the electrons will possess many times this energy enabling them to cause excitation and ionization.

The atomic gas, on the other hand, will only be at a temperature of a few hundred degrees but the energy transfer between the electron and atomic gas becomes more significant at a higher pressure where the collision rate increases dramatically. A point is then reached where the temperatures of the two gases reach an average value, and a local thermal equilibrium is then established (Figure 3.37).

3.3.2 Line Broadening

The pressure increase also affects the width of the lines. In a low-pressure discharge, the width of the atomic lines is Doppler dominated. This means that the lines have a finite width due to their thermal motion, but at a pressure of more than a few atmospheres the lines are significantly broadened due to the perturbation of the energy level by other atoms that are close by. This perturbation is even more pronounced when the interaction is between identical atoms. An example of this is the high-pressure sodium lamp where the color rendering properties of the lamp are greatly improved due to the resonant broadening that occurs, as will be noted later.

Table 3.10. General Comparisons of Low- and High-Pressure Discharge Lamps

Low-Pressure Discharge Tube	High-Pressure Discharge Tube
Electron temperature = 11,000 K	Electron temperature = 5000 K
Gas temperature = 350 K	Gas temperature = 5000 K
Mainly emissions of the resonance lines	Emission lines from higher energy states
Usually long tube	Usually short tube
Low brightness	High brightness
Pressure lower than atmospheric	Pressure higher than atmospheric
Power less than 1 kW	Power of up to several kW
Tube made usually of soft glass	Tube usually made of quartz or PCA

3.4   High-Pressure Discharge Lamps

In cases where high brightness is required with relatively high efficiency and good color rendering, high-pressure discharge lamps can be used. The lamps operate using the same principle of electric discharge in a gas or vapor that results in excitation of the active medium atoms and the subsequent emission of radiation. As the title of this group of lamps indicates, the difference is that the gas or vapor pressure is much higher (several atmospheres).

In a high-pressure discharge tube the wattage per discharge centimeter (power density) is much higher than that of a low-pressure discharge tube so the number of collisions between electrons and atoms is also much higher. The higher number of collisions results in an increase of the gas temperature due to a transfer of kinetic energy and consequently to increased gas pressure. The electrodes emit electrons due to their high temperature (thermionic emission) without the aid of an electric field and the emissions of the active medium atoms are not limited to the resonance lines. Table 3.10 summarizes the differences between low- and high-pressure discharges.

The high-pressure discharge lamp consumes larger amounts of energy than a low-pressure one and is characterized by a far greater efficiency than an incandescent lamp of equal power. This category of lamps is known as HID (high-intensity discharge) lamps. Just like the low-pressure discharge lamps, they require a ballast to keep the intensity of the current to a desired value. There are many types of gear and their choice depends on the power consumption of the lamp and the external operating conditions.

A high-pressure discharge starts with pulses of high voltage applied between the two electrodes while in some tubes a third electrode is placed adjacent to one of the others so that a smaller discharge causes ionization and makes the gas or vapor conductive. The buffer gas also plays a role in creating a discharge and starting the lamp. Because of their high brightness and efficacy, these lamps are employed for lighting large spaces such as warehouses, stadiums, public places, parking lots, and roads. Another application where such high-pressure discharge lamps are encountered is automotive lighting where they replace conventional incandescent lamps as headlights.

3.4.1   High-Pressure Mercury Vapor Discharges

Mercury is used as the active medium, not only in fluorescent lamps, but also in high-pressure discharges. The glass of the bulb is quartz, which is more

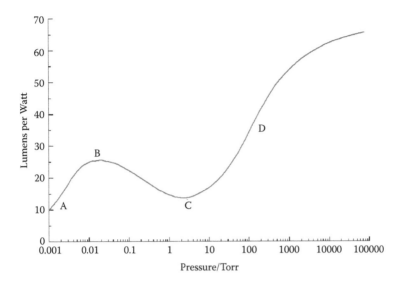

Figure 3.38. The luminous efficiency of a gas discharge varies with the Hg vapor fill pressure.

resistant to high temperatures and pressure, and is also enclosed within another larger glass body, which is under a vacuum. Due to the higher number of electron impacts, the atoms of mercury are excited to high-energy states leading to emission spectra with visible lines of higher intensity than those found in the spectrum of a fluorescent lamp.

Figure 3.38 shows the dependence on the efficacy of mercury vapor pressure. At point A, the efficacy is very low since the conditions there favor the production of the ultraviolet resonant lines. With increasing pressure, however, absorption of the resonant lines causes an increase in the population of higher energy levels. Since these levels include the visible mercury lines, the increase in their population results in an improvement of the efficacy of the discharge. The decrease in the efficacy beyond point B is due to the fact that elastic collisions between electrons and mercury atoms increase with increasing pressure, so more energy is lost due to elastic collisions and, thus, the efficacy decreases; at the same time, the gas temperature rises resulting in transfer of the energy input to the tube walls due to thermal conduction. Although the gas temperature increases with increasing pressure, the electron temperature decreases as a result of the larger number of collisions.

When the pressure exceeds point C, the gas temperature has become so high that thermal excitation of mercury atoms is possible, so the luminous efficiency of the discharge increases again. So, the "low-pressure" and "high-pressure" discharges differ considerably from each other in their fundamental mechanism, because the former consists mostly of resonant transitions while the latter has transitions between the upper energy levels. Perhaps a better distinction between these two kinds of discharges can be described as "nonthermal" and "thermal."

Figure 3.39. High-pressure mercury vapor discharge lamp (Iwasaki H2000B Clear Hg 2000 W).

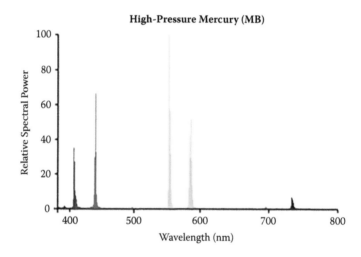

Figure 3.40. Emission spectrum of a high-pressure mercury vapor discharge lamp.

Once again we deal with a line spectrum and not a continuous one so the final result is sometimes formed using a fluorescent powder that converts any UV radiation emitted into visible light. The phosphor is coated on the inside wall of the outer tube, which serves as a thermal insulator. In general, the spectrum is deficient in emissions in the red part and the characteristics of such a lamp are defined by the choice of phosphor. Some indicative values for the lamp characteristics are: luminous efficacy of 60 lm/W, color rendering index of 15–55, color temperature at around 3000–4000 K, and an average lifetime of about 15,000 hours. Figure 3.39 through Figure 3.43 provide examples of commercial high-pressure mercury vapor lamps and typical emission spectra while Tables 3.11 and 3.12 list some general characteristics.

Figure 3.41. High-pressure mercury vapor discharge lamp (Philips HPL + YAG + Yttrium Vanadate coating 80 W).

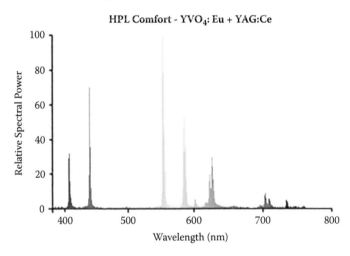

Figure 3.42. Emission spectrum of a high-pressure mercury vapor discharge lamp with phosphor.

A variation of the high-pressure mercury vapor lamp is the super-high-pressure mercury lamp (200 Atm) that does not employ a secondary outer tube and is used in projectors for image and video displays (Figure 3.44; spectrum in Figure 3.45; and characteristics in Table 3.13). In this case, the mercury emission lines are strongly broadened due to the high pressure (pressure broadening) and cover the entire visible spectrum, resulting in a high-intensity white light with no need for fluorescent powders.

Yet another variation is the hybrid lamp that combines the mercury vapor discharge and incandescence (Figure 3.46; spectrum in Figure 3.47; and characteristics in Table 3.14). The lamp emits a warm white light with low luminous efficacy

Figure 3.43. Electrode of a high-pressure mercury vapor discharge lamp.

Table 3.11. Table with Indicative Values and Information about the General Characteristics of the High-Pressure Mercury Vapor Lamp

Gas fill	Inner: Ar|Hg	Outer: N$_2$
Luminous flux	120,000 lm (100 h)	
Efficacy	60 lm/W (100 h)	
Color temperature and CRI	CCT: 5700 K	CRI: Ra = 15
Average lifetime	10,000 h	

Table 3.12. Table with Indicative Values and Information about the General Characteristics of the High-Pressure Mercury Vapor Lamp

Gas fill	Inner: Ar|Hg	Outer: N$_2$
Luminous flux	4000 lm (100 h)	
Efficacy	50 lm/W (100 h)	
Color temperature and CRI	CCT: 3500 K	CRI: Ra = 57
Color coordinates	CCx: 0.40	CCy: 0.38
Average lifetime	28,000 h	

and poor color rendering. The incandescent filament that contributes to the light emissions (in the red part) also plays the role of current intensity regulator as a resistance, that is, the lamp is self-ballasted.

3.4.2 High-Pressure Sodium Vapor Discharges

The high-pressure sodium vapor discharge tubes, in fact, employ a mix of sodium and mercury, resulting in the emission of radiation with greater coverage of the visible spectrum. Another difference in high-pressure mercury lamps is the material used for the construction of the inner bulb that contains the metallic active media. Quartz, which is used to manufacture high-pressure mercury lamps; and boron glass, which is used to manufacture low-pressure sodium lamps, cannot

Figure 3.44. Super-high-pressure mercury vapor discharge lamp (Philips UHP LCD Video Projector 120 W).

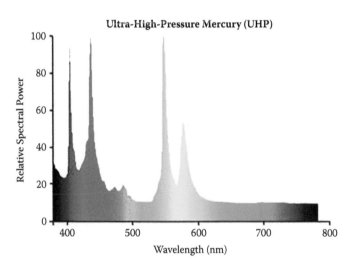

Figure 3.45. Emission spectrum of a super-high-pressure mercury vapor discharge lamp.

Table 3.13. Table with Indicative Values and Information about the General Characteristics of the Super-High-Pressure Mercury Vapor Lamp

Gas fill	Inner: Ar-Br|Hg	
Luminous flux	7000 lm (100 h)	
Efficacy	58 lm/W (100 h)	
Color temperature and CRI	CCT: 7600 K	CRI: Ra = 57
Color coordinates	CCx: 0.298	CCy: 0.311
Average lifetime	6000 h	

Figure 3.46. Hybrid high-pressure mercury lamp (Philips ML Blended Self-Ballasted 160 W).

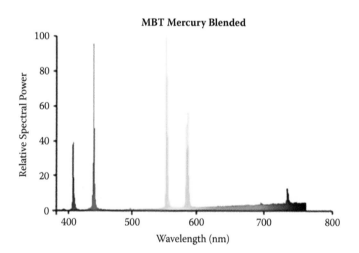

Figure 3.47. Emission spectrum of a hybrid mercury lamp.

Table 3.14. Table with Indicative Values and Information about the General Characteristics of the Hybrid Mercury Lamp

Gas fill	Inner: Ar|Hg	Outer: N_2
Luminous flux	3000 lm (100 h)	
Efficacy	18 lm/W (100 h)	
Color temperature and CRI	CCT: 3800 K	CRI: Ra = 60
Color coordinates	CCx: 0.349	CCy: 0.380
Average lifetime	6000 h	

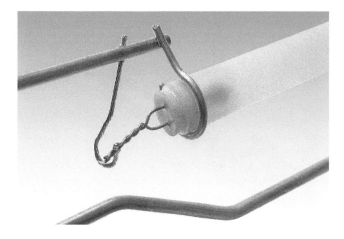

Figure 3.48. Polycrystalline alumina burner (PCA).

withstand the corrosivity of sodium under high temperatures and pressure so a crystalline material of aluminum oxide (polycrystalline alumina known as *PCA*) is used. Lamps made from this material are also known as *ceramic burners*, see Figure 3.48 for an example.

Due to their continuous emission spectrum, the high-pressure sodium vapor lamps have a good color rendering index of up to 85 and the luminous efficacy is quite high and exceeds 100 lm/W (taking into account the sensitivity curve of the eye, the efficacy reaches 150 lm/W). The color temperature is around 2000–2500 K, which is a warm white, and the average life of the bulb can be up to 30,000 hours. Their applications are the same as the high-pressure mercury vapor lamps except that the high-pressure sodium lamps offer better color rendering. Figure 3.49, Figure 3.51, Figure 3.53, and Figure 3.55 show commercial high-pressure sodium vapor discharge lamps while their respective typical spectra are shown in Figure 3.50, Figure 3.52, Figure 3.54, and Figure 3.56. Tables 3.15 through 3.18 list some of their general lamp characteristics. Table 3.19 summarizes the different high-pressure sodium lamp designs.

3.4.3 Metal Halide Discharge Lamps

A further development in the high-pressure mercury or sodium vapor discharge lamps is the addition of more active media in order to tailor the emission spectrum and have control over the color rendering index or the color temperature. The addition of other pure metals known for their number of emission lines in the visible spectrum would not have the desired effect because of their very low partial pressure, even at temperatures encountered in high-pressure tubes.

A solution for overcoming the problem of the low-vapor pressure of pure metals is the addition of their halide compounds, which are easier to vaporize. Once in the gas phase, the compounds are dissociated due to electron collisions and the freed metallic atoms become excited. The excitation is followed by the relaxation of the atoms and the extra energy is released in the form of emitted

Figure 3.49. High-pressure sodium vapor discharge lamp (Reflux with a high special side reflector 250 W).

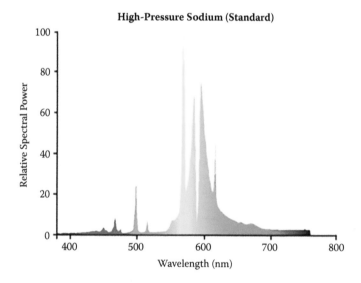

Figure 3.50. Emission spectrum of a high-pressure sodium vapor discharge lamp.

electromagnetic waves with a frequency depending on the metal of choice and its signature energy-level structure.

$$MX_n \leftrightarrow M + nX$$

$$M + e^- \rightarrow M^*$$

$$\rightarrow M^+ + 2\,e^-$$

$$M^* \rightarrow M + h\nu$$

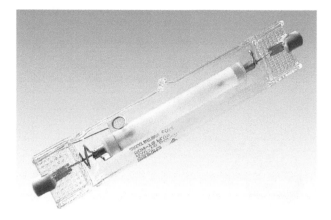

Figure 3.51. High-pressure sodium vapor discharge lamp (SON-TS Vialox Super Vacuum Linear 150 W).

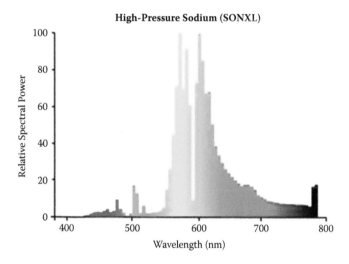

Figure 3.52. Emission spectrum of a high-pressure sodium vapor discharge lamp.

where M is the metal, X is the halogen, n is the stoichiometric number, h is Planck's constant, and v is the frequency of the emitted wave.

The combination of emitted radiation from various metallic elements allows the development of lamps with a variety of features. These lamps, which are called *metal halide lamps*, have a color rendering index up to 90 and their color temperature starts at 3000 K and reaches 20,000 K. The luminous efficacy is generally high with over 110 lm/W for some lamps and the average lifetime reaches 20,000 hours. Figures 3.57 and 3.58, Figure 3.60, Figure 3.62, and Figure 3.64 provide some examples of commercial metal halide lamps. Figure 3.59, Figure 3.61, Figure 3.63, and Figure 3.65 illustrate their respective emission spectra, and Table 3.20 through Table 3.23 list some of the metal halide characteristics.

Figure 3.53. High-pressure sodium vapor discharge lamp (Philips SON City Beautification 150 W).

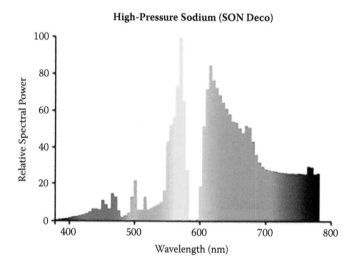

Figure 3.54. Emission spectrum of a high-pressure sodium vapor discharge lamp.

The construction of this type of lamp is similar to that of other high-pressure lamps with the internal burner typically made of crystalline aluminum oxide (PCA) due to the use of corrosive elements and compounds. The outer tube is not coated with phosphors, but acts as a filter for ultraviolet radiation emitted by mercury and often there is a coating to diffuse the light emitted.

3.4.4 High-Pressure Xenon Gas Discharges

The last example in this group of high-pressure lamps is the high-pressure xenon discharge tube. The burner (tube) is made of quartz and the electrodes are made of tungsten with traces of thorium. The emitted radiation covers the entire visible

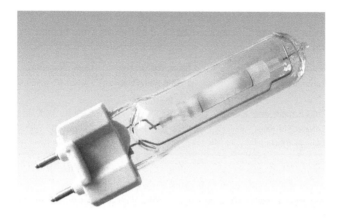

Figure 3.55. High-pressure sodium vapor discharge lamp (Philips SDW-TG Mini White 100 W).

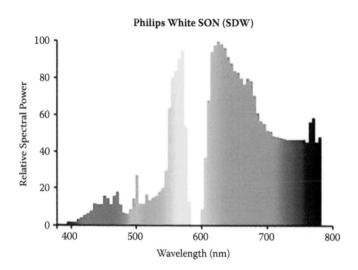

Figure 3.56. Emission spectrum of a high-pressure sodium vapor discharge lamp.

Table 3.15. Table with Indicative Values and Information about the General Characteristics of the High-Pressure Sodium Vapor Lamp

Gas fill	Inner: XeI Na, Hg	Outer: Vacuum
Luminous flux	26,000 lm (100 h)	
Efficacy	104 lm/W (100 h)	
Color temperature and CRI	CCT: 1900 K	CRI: Ra = 25
Color coordinates	CCx: 0.542	CCy: 0.415
Average lifetime	24,000 h	

Table 3.16. Table with Indicative Values and Information about the General Characteristics of the High-Pressure Sodium Vapor Lamp

Gas fill	Inner: Xe|Na, Hg	Outer: Vacuum
Luminous flux	15,000 lm (100 h)	
Efficacy	100 lm/W (100 h)	
Color temperature and CRI	CCT: 2000 K	CRI: Ra = 25
Color coordinates	CCx: 0.530	CCy: 0.430
Average lifetime	12,000 h	

Table 3.17. Table with Indicative Values and Information about the General Characteristics of the High-Pressure Sodium Vapor Lamp

Gas fill	Inner: Xe|Na, Hg	Outer: Vacuum
Luminous flux	7000 lm (100 h)	
Efficacy	48 lm/W (100 h)	
Color temperature and CRI	CCT: 2500 K	CRI: Ra = >80
Average lifetime	8000 h	

Table 3.18. Table with Indicative Values and Information about the General Characteristics of the High-Pressure Sodium Vapor Lamp

Gas fill	Inner: Na, Hg/Xe	Outer: Vacuum
Luminous flux	4800 lm (100 h)	
Efficacy	48 lm/W (100 h)	
Color temperature and CRI	CCT: 2500 K	CRI: Ra = 83
Color coordinates	CCx: 0.470	CCy: 0.406
Average lifetime	10,000 h	

Table 3.19. Characteristics of High-Pressure Sodium Vapor under Different Design Elements

Design Element	Efficacy lm/W	Color Rendering Index	Color Temperature/K	Disadvantages
Simple	60–130	20–25	2000	
Increased xenon pressure	80–150	20–25	2000	Difficult to start
Increased sodium vapor pressure	60–90	60	2200	Lower efficacy and shorter lifetime
Further increase of sodium vapor pressure	50–60	85	2500	Lower efficacy and shorter lifetime
Electronic ballast for pulsed operation	50–60	85	2600–3000	Lower efficacy and shorter lifetime Special operating gear

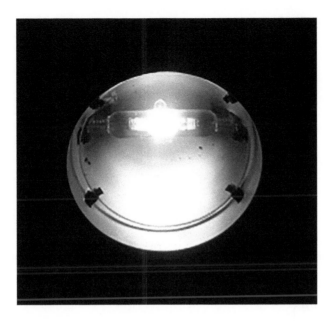

Figure 3.57. Emission of light by a metal halide lamp.

spectrum, but special attention is required because of the very high pressure and ultraviolet radiation that is emitted, which penetrates the quartz without loss of intensity. The lamps are of low luminous efficacy, but are used wherever white light is wanted to simulate daylight.

Several sizes of this lamp are available on the market—they are divided into small and large xenon arc tubes ranging in power from a few watts to many kilowatts. The term *arc* refers to high-pressure discharges that are due to the exact shape given to the plasma by the thermal buoyancy. The lamps have a high luminous flux, which reaches 200,000 lm under pulsed conditions. Figure 3.66 shows a typical high-pressure xenon discharge lamp.

In a variation of xenon lamps, mercury is added, giving the white light a blue tone, increasing the intensity of UV radiation. Therefore, xenon-mercury lamps are used in applications where ultraviolet radiation is desired for sterilization or ozone formation.

A marriage of technologies, those of metal halide and high-pressure xenon lamps, can be found in automotive lights (Figure 3.67) as mentioned earlier, but they have become known as xenon lamps although xenon serves as the buffer gas and only emits during the first warm-up minutes before the metal halides vaporize. These lamps operate at 40 watts of power and emit about 3000 lm.

3.4.5 Carbon Arc Lamps

The carbon arc lamp is a lamp that combines a discharge with incandescence. The starting of the lamp is achieved by bringing in contact two carbon electrodes so that a discharge is created with a relatively low voltage application. The electrodes are gradually moved away from each other and the electricity current already

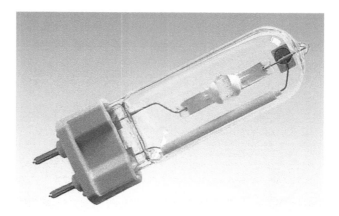

Figure 3.58. Metal halide lamp (Philips MHN-TC Single-Ended Quartz 35 W).

MHN-TC Dysprosium-Thallium-Sodium

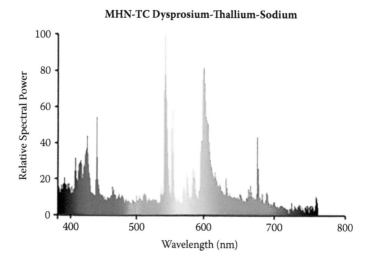

Figure 3.59. Emission spectrum of a high-pressure metal halide discharge lamp.

flowing maintains the arc. The edges of the electrodes are at a high temperature so they too emit light due to incandescence. Because of their vaporization, the electrodes must periodically be brought closer to each other so that the distance between them remains constant and the arc is maintained. Apart from some specific applications such as video projector screenings, the carbon arc lamps have essentially been retired and replaced by xenon arcs.

3.5   Induction Lamps

Unlike all other electrical lamps that use electrodes to couple energy (electricity) to the tube, there is a class of lamps where the power is transferred to the gas or

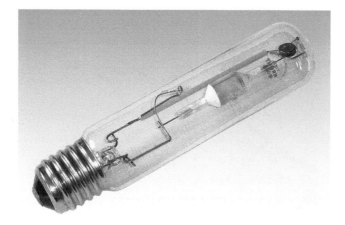

Figure 3.60. Metal halide lamp (Thorn Kolorarc MBI-T Tri-Band Chemistry 250 W).

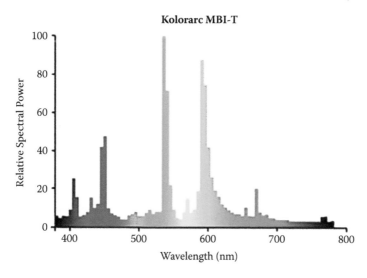

Figure 3.61. Emission spectrum of a high-pressure metal halide discharge lamp.

vapor inductively. The great advantage of induction lamps is that the absence of electrodes gives them a much longer lifetime. In induction lamps, one has the possibility to use a greater variety of materials as active media than those that would be compatible with just electrode materials in conventional tubes.

3.5.1   Low-Pressure Mercury Vapor Induction Lamps

Let's first consider the case of the low-pressure mercury vapor induction lamp. The operating principle is the same as with other fluorescent low-pressure mercury lamps, as mercury atoms are excited through electron impacts and after relaxation of the atoms to their ground state they emit radiation with the ultra-violet (254 nm) resonance line being the main one. The ultraviolet light is then

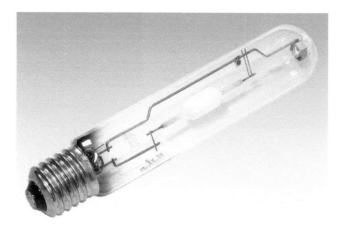

Figure 3.62. Metal halide lamp (GE CMH-TT 250 W/830).

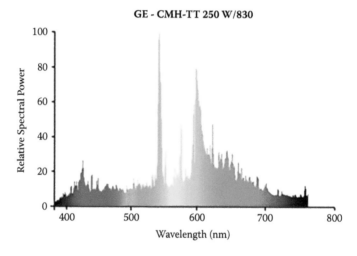

Figure 3.63. Emission spectrum of a high-pressure metal halide discharge lamp.

converted to visible light by a phosphor coated on the inside walls of the lamp. The difference here is that there are no electrodes to provide the flow of electrons. Instead, the free electrons in the lamp are set in motion by an electromagnetic field created by an antenna (induction) mounted adjacent to the walls of the lamp. Different companies have presented different setups for this antenna mounting and some examples can be seen in Figure 3.68 through Figure 3.70 presenting different aspects of the same commercial product, and Figure 3.72 shows another brand. The emission respective spectra are illustrated in Figures 3.71 and 3.73, while the general characteristics of these two induction mercury lamps are listed in Tables 3.24 and 3.25. The electronic ballast generates this high-frequency electromagnetic field that is transferred through the antenna to the gas. Some common frequencies are 13.6 MHz and 2.65 MHz, that is, in the range of microwaves.

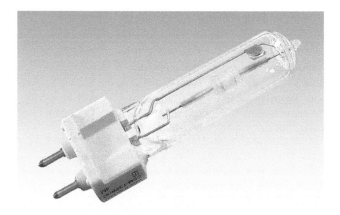

Figure 3.64. Metal halide lamp (Philips CDM-T Single-Ended 35 W).

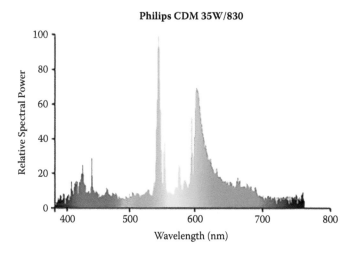

Figure 3.65. Emission spectrum of a high-pressure metal halide discharge lamp.

While the average lifetime of electroded fluorescent lamps depends on the average lifetime of their electrodes, the lifetime of the induction fluorescent lamp is defined by the average life of the operating gear, which exceeds 30,000 hours.

3.5.2 Sulfur Lamps

The sulfur lamp is an example of where the inductive operating mode allows the use of an active medium that could not have been used in a discharge tube with electrodes. Here, again, a magnetron creates an electromagnetic field at the frequency of 2.65 MHz, which sets in motion the free electrons and raises the temperature of the lamp and consequently the vapor pressure of the sulfur (5 Atm). Due to the high temperature and pressure, the spherical bulb is made of quartz, which can withstand the conditions (Figure 3.74). The luminous flux of a

Table 3.20. Table with Indicative Values and Information about the General Characteristics of the Metal Halide Lamp

Gas fill	Inner: Ar/Dy, Tl, Na	Outer: Vacuum
Luminous flux	2600 lm (100 h)	
Efficacy	67 lm/W (100 h)	
Color temperature and CRI	CCT: 3800 K	CRI: Ra = 75
Color coordinates	CCx: 0.385	CCy: 0.368
Average lifetime	6000 h	

Table 3.21. Table with Indicative Values and Information about the General Characteristics of the Metal Halide Lamp

Gas fill	Inner: Ar|Hg-In (Na, Tl, Hg) Ix	Outer: N_2
Luminous flux	21,000 lm (100 h)	
Efficacy	84 lm/W (100 h)	
Color temperature and CRI	CCT: 4200 K	CRI: Ra = 70
Average lifetime	6000 h	

Table 3.22. Table with Indicative Values and Information about the General Characteristics of the Metal Halide Lamp

Gas fill	Inner: Ar/Hg	Outer: Vacuum
Luminous flux	25,000 lm (100 h)	
Efficacy	100 lm/W (100 h)	
Color temperature and CRI	CCT: 3150 K	CRI: Ra = 85
Color coordinates	CCx: 0.424	CCx: 0.395
Average lifetime	20,000 h	

Table 3.23. Table with Indicative Values and Information about the General Characteristics of the Metal Halide Lamp

Gas fill	Inner: Ar-Kr_{85}|Dy-Ho-Tm-Tl-Na-Ix	Outer: Vacuum
Luminous flux	3300 lm (100 h)	
Efficacy	95 lm/W (100 h)	
Color temperature and CRI	CCT: 3000 K	CRI: Ra = 81
Color coordinates	CCx: 0.428	CCy: 0.397
Average lifetime	6000 h	

sulfur lamp is quite high (see Table 3.26) compared to other lamp technologies, and lighting a building's interior is possible by directing the emitted light of one lamp to the desired spaces through optical tubes. The spectrum of such a lamp is shown in Figure 3.75. The emissions are the result of excitations not of sulfur atoms but of sulfur molecules (S2), therefore the continuous spectrum covers the entire visible region. The luminous efficacy of these lamps exceeds 100 lm/W and the consuming power is on the order of kilowatts.

The color temperature of this light source is 6000 K and the color rendering index is around 80. The spectrum and therefore some characteristics of this white

Figure 3.66. High-pressure xenon lamp.

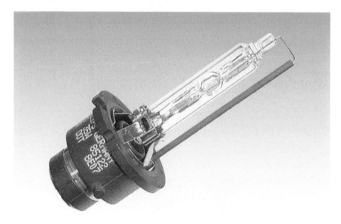

Figure 3.67. Metal halide/xenon lamps for car headlights.

light source, such as the color rendering index and the color temperature, can be affected or tailored by the addition of other chemicals (CaBr2, LiI, NaI). The average lamp lifetime is defined by the life of the magnetron and reaches 20,000 hours.

Although not widely used, the sulfur lamp is a light source with very high luminous flux but without the emission of ultraviolet or infrared radiation. It is usually employed for lighting large areas either as it is or with the use of light pipes that distribute the light uniformly over their length and carry the light to many places thus eliminating the need for other lamps.

Due to the electromagnetic waves emitted with both types of induction lamps, special filters are placed around the bulb to reduce interference to other receivers of this wavelength.

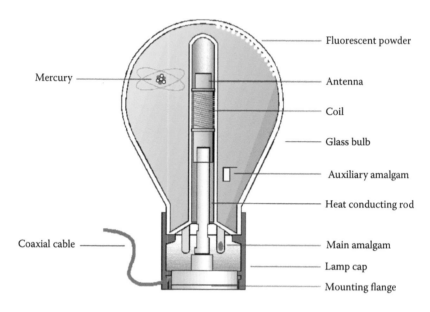

- Fluorescent powder
- Antenna
- Coil
- Glass bulb
- Auxiliary amalgam
- Heat conducting rod
- Main amalgam
- Lamp cap
- Mounting flange

Mercury

Coaxial cable

Figure 3.68. Anatomy of an induction low-pressure mercury vapor lamp.

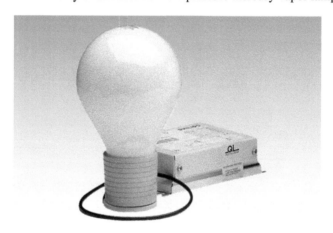

Figure 3.69. An induction low-pressure mercury vapor lamp (Philips QL Electrodeless Induction System 85 W).

3.6   Pulsed Operation

The use of a pulsed current has opened new horizons in the field of research and development for novel light source technologies and some of the new ideas have already taken root in the production of new products.

For a discharge lamp of a given power input, the change from a continuous flow of charge (regardless of whether it is unipolar or bipolar) to a pulsed operation (burst of charge flow) means that the average electron energy shifts to higher values since the applied voltage and current reaches for short periods much

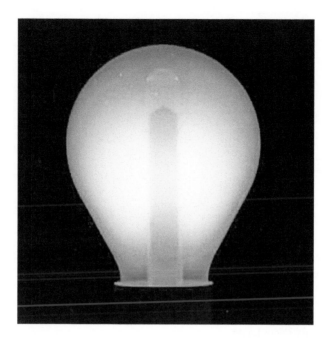

Figure 3.70. Photograph of an induction mercury lamp without the phosphor coating.

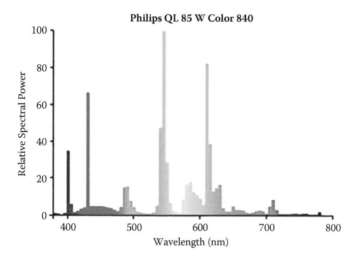

Figure 3.71. Emission spectrum of a low-pressure induction mercury vapor lamp.

higher values in order for the discharge to be maintained and the average power to remain the same. This shift in average electron energy can significantly affect the emission spectrum of the lamp. Figure 3.76 shows the different waveforms between the two modes of operation.

The case of mercury that is being employed in the majority of commercial discharge lamps for general lighting is examined next.

Figure 3.72. Induction low-pressure mercury vapor lamp (Osram Endura Inductively Coupled Electrodeless 100 W).

Table 3.24. Table with Indicative Values and Information about the General Characteristics of the Low-Pressure Induction (Electrodeless) Mercury Vapor Lamp

Gas fill	Ar-Kr-Ne|Hg Amalgam	
Luminous flux	6000 lm (100 h)	4200 lm (60,000 h)
Efficacy	71 lm/W (100 h)	49 lm/W (12,000 h)
Color temperature and CRI	CCT: 4000 K	CRI: Ra = 80
Color coordinates	CCx: 0.390	CCy: 0.390
Average lifetime	100,000 h	

Table 3.25. Table with Indicative Values and Information about the General Characteristics of the Low-Pressure Induction (Electrodeless) Mercury Vapor Lamp

Gas fill	Ar-Ne-Kr/Hg	
Luminous flux	8000 lm (100 h)	5600 lm (60,000 h)
Efficacy	80 lm/W (100 h)	56 lm/W (60,000 h)
Color temperature and CRI	CCT: 3000 K	CRI: Ra = 80
Average lifetime	60,000 h	

The emission spectrum of a low-pressure mercury vapor discharge tube under steady- or pseudo-steady-state operation (DC or AC current) is known (Figure 3.77) and the dominant resonance emission line at 254 nm is the one that excites the phosphor on the tube walls and converts to most of the visible light output.

Only a small fraction of electrons on the high-energy tail can induce, via collisional excitation, atomic transitions, which could lead to the emission of near-UV and visible lines and also to ionization, which is essential for the maintenance of the plasma.

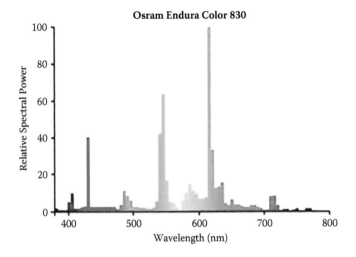

Figure 3.73. Emission spectrum of a low-pressure induction mercury vapor lamp.

Figure 3.74. TUE electrodeless high-pressure sulfur microwave lamp 1000 W.

Table 3.26. Table with Indicative Values and Information about the General Characteristics of the High-Pressure Induction (Electrodeless) Sulfur Vapor Lamp

Gas fill	S_2–26 mg (5 bar) Ar-75 Torr	
Luminous flux	130,000 lm (100 h)	
Efficacy	130 lm/W (100 h)	95 lm/W system
Color temperature and CRI	CCT: 6000 K	CRI: Ra = 79
Average lifetime	60,000 h (lamp)	20,000 h (magnetron)

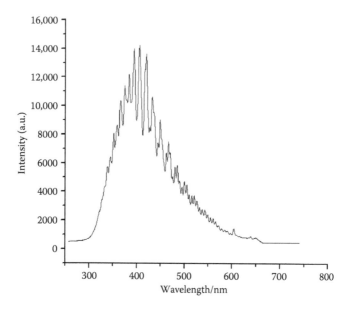

Figure 3.75. Emission spectrum of a high-pressure sulfur induction lamp.

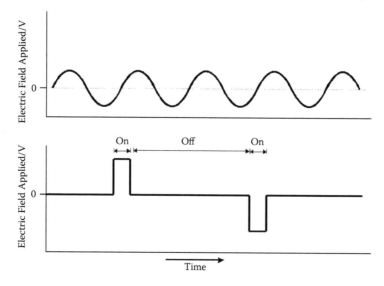

Figure 3.76. Waveforms of applied voltage for continuous (AC in this example) and pulsed operation of a discharge tube.

In the pulse operation of a discharge tube (pulses of a few microseconds long and a frequency of a few kHz), the mean energy of electrons is increased therefore increasing the ionization degree of the mercury atoms and their excitation to higher energy states than that of resonance. The spectrum shows an enhancement of the near-UV (313 and 365 nm) and visible lines at the expense of the resonance

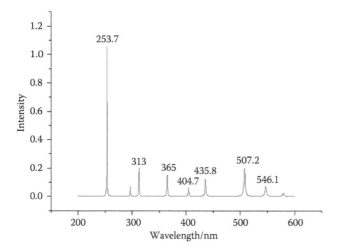

Figure 3.77. Emission spectrum of a low-pressure mercury vapor discharge tube under steady-state operation.

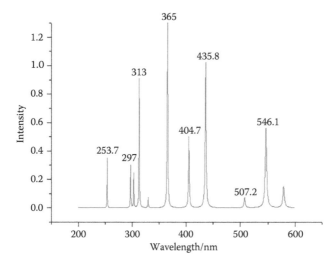

Figure 3.78. Emission spectrum of a low-pressure mercury vapor discharge tube under pulsed operation. Instrument sensitivity must also be taken into account when considering absolute intensities. This graph represents relative intensities and how the near-UV and visible lines were enhanced at the expense of the resonant UV line.

line [3] as illustrated in Figure 3.78. The relative intensities can be controlled by controlling the duration or frequency of the pulses and consequently the applied voltage and current waveforms. One possibility arising from this effect is the use of lines close or in the visible part of the spectrum for phosphor excitation in order to reduce or eliminate Stokes' losses.

An important point here is that during pulsed operation and as the ionization degree increases, the emission of radiation continues in between pulses in the form of afterglow emissions. This afterglow regime is not the same for all emission lines and as expected, the lines that exhibit the stronger afterglow emissions are the ones that are mostly enhanced under this mode of radiation.

Details of the electron energies distribution function (EEDF) are very important for the development of models that can predict the behavior of such plasmas, but it is generally assumed that the distribution becomes Maxwellian sometime after the cessation of the pulse.

Another application could be the use of a mixture of phosphors with each being excited by a different mercury line (for example, the use of a phosphor excited by the 254 nm line which dominates during steady-state operation in combination with a second phosphor that is excited by the 365 nm line that is prominent under pulsing). The alternation of the operating modes or the control of the pulse duration and frequency can allow for the tailoring of the emission spectrum and therefore the control and variation of the color temperature or rendering index of the source.

The pulsed operation mode was employed in another pioneering work where for the first time the focus was medium pressure mercury vapor discharge lamps. The increase of mercury pressure to regimes of a few Torr (1–100) and the pulse operation resulted in a white light source without the need for phosphors. Pulsing and increasing the pressure once again caused an enhancement of the near UV and visible lines at the expense of the resonance line but under these conditions the yellow double line of mercury at 577/579 nm was particularly enhanced (Figure 3.79). This increase in intensity of the yellow line is what causes the

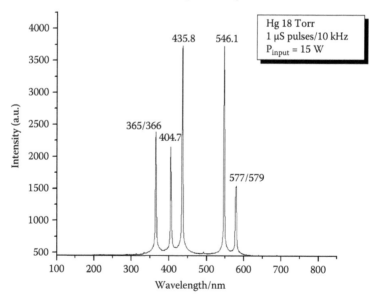

Figure 3.79. Emission spectrum of a medium-pressure mercury discharge tube.

Figure 3.80. A discharge through pure mercury vapor at a pressure of 18 Torr without a buffer gas appears white to the naked eye.

source to appear white (Figure 3.80) as yellow activates the same eye sensors as red, and in combination with the other green and blue lines, creates the effect of white light. It is only under the pulsed medium pressure conditions that the yellow line makes an impact [4].

The adoption of the pulse operating mode will probably not be restricted to discharges of different pressure regimes as well as in the use of other technologies. LEDs and incandescent lamps could probably be pulsed in order to control the temperature (of the junction or filament) or the response of the human eye.

3.7   Alternative Discharge Light Sources

The fact is that the use of pure elements as active media poses the limitation of insufficient vapor pressure in most cases with few exceptions that have already been tested and are well known. Elements like the rare gases, mercury and sodium, have already been the base for a number of products while other elements with useful emission properties can only be used in molecular form and in high-pressure lamps where the temperatures reach high enough values for them to be vaporized and dissociated to a useful degree.

A paper by Zissis and Kitsinelis [5] lists most of the attempts made in recent years to find alternatives to mercury discharge lamps, and experimentation from various groups thus far includes:

- Barium in low-pressure regimes (visible resonance line)

- Nitrogen as a near-UV emitter, oxides of metals (electrodeless configurations to avoid oxygen's reactions with lamp parts)

- Zinc in high-pressure regimes

- Hydroxyl (OH) as a near-UV emitter in low-pressure regimes with rare gas backgrounds

- Cyanogen radical (CN) as a near-UV emitter in low-pressure regimes

- Carbon monoxide (CO) in low-pressure regimes as a visible continuous source

- Sodium iodide (NaI), scandium iodide (ScI), indium monobromide (InBr), and other metal halide salts under microwave excitation

In all the above cases, the efficacies were lower than those of existing discharge lamps, and, in some cases, the necessary changes in lamp design and technology constitute further barriers for the development of such solutions.

Especially for the low-pressure cases, all recent attempts to replace mercury have focused either on atomic species alone or included the use of molecular compounds but examined the vapor pressure and the emission properties of the molecular candidates and not those of the constituent atoms in the case of atomization. The belief that significant atomization occurs and that atoms and diatomics dominate the emission properties of low-pressure discharges regardless of the molecules employed stems from previous experience on mercury-free, low-pressure discharges [6]. Bearing these extra elements in mind (up to two atoms per species and significant atomization) the author approached the issue in the past by redesigning the strategy and selection rules [7].

The first criterion is that the new candidates must be efficient emitters in the range of 380–580 nm. The reason for selecting this particular range is the need to eliminate any Stokes' losses that would be associated with photons the human eye cannot detect and would need conversion with the use of a phosphor. On the short-wavelength side, the eye sensitivity and therefore the visible range starts at around 360–380 nm. On the long-wavelength side, there was no real need to include wavelengths beyond 580 nm as sodium is already a strong emitter, and just above that limit, the sensitivity of the eye drops and in any case the photons on that part of the spectrum cannot be converted with the use of phosphor. On the other hand, strong emissions in the blue region could be partly converted in order to cover larger regions of the visible spectrum.

The second requirement is that the emitter has to be in vapor phase or easily evaporable. Of course, the selected emitter should not be aggressive for the burner envelope and electrode materials. Toxic or radioactive substances also cannot be considered for environmental purposes.

The search should focus on diatomic species (so as to overcome at least partly the vapor pressure limitations of the atomic elements), with at least one of the atoms being an element with useful emissions due to the fact that even with molecules in the gas phase, significant atomization occurs and the atomic emissions contribute significantly to the overall radiated power output. Finally, one should search for parent molecules if the diatomic species are not freely available or stable in their form [8,9].

Any compound chosen must also be in the lowest possible oxidation state that still vaporizes to the desired degree in order to avoid stoichiometrically driven condensation. What is meant by this statement can be explained with an example. Consider a metal M that vaporizes to the desired degree when in the form of MX_4 where X is a halogen or an oxygen atom (or a combination of both as in some oxohalide compounds) but condenses when MX_2 forms, then in the plasma phase of a lamp that starts with MX_4, all metal will in time condense as it forms MX_2 and X_2 in the gas phase (such as molecular oxygen or halogen). On the other hand, if the products of the plasma reactions do not form a stable species with very low vapor pressure under those conditions, this problem can be avoided.

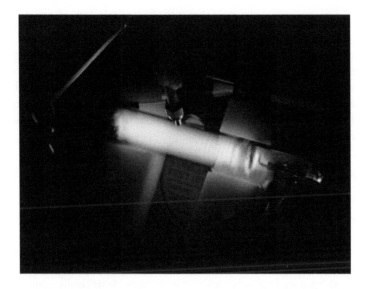

Figure 3.81. Photograph of a low-pressure metal halide lamp (AlCl3), which has been proposed as an alternative to mercury-filled, low-pressure discharge lamps.

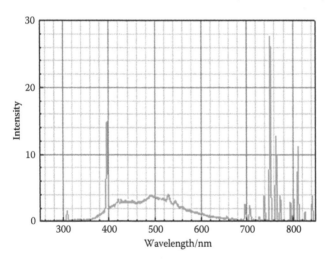

Figure 3.82. Emission spectrum of a low-pressure metal halide lamp (AlCl3), which has been proposed as an alternative to mercury-filled, low-pressure discharge lamps.

The author has developed an $AlCl_3$ low-pressure lamp (Figure 3.81; spectrum in Figure 3.82) [7] after applying a proposed strategy and selection rules. A large molecular band exists between 360 and 650 nm. Overall, the spectrum containing both atomic and molecular emissions justifies the strategy and rules applied.

Having identified promising candidates as active media, the proposed technology of molecular low-pressure, mercury-free gas discharges essentially requires the use of the principles of metal halide lamps where a variety of spectra can be produced by introducing a mixture of metal compounds and the infrared reflecting coating used in incandescent and sodium lamps in order to provide thermal insulation and maintain the temperature as high as possible. The use of these coatings could provide the temperature conditions for metal compounds to be vaporized even in low-pressure and low-power discharge tubes while the use of an electrodeless tube can solve the problem of the corrosive properties of these metal compounds (see the sulfur lamp example in Section 3.5.2).

3.8   Microdischarge Properties for the Development of Light Sources

Microdischarges emerged as a new front in the research for nonequilibrium plasma, introducing a wide range of possible applications, most significantly in the field of light sources, biomedical applications, and nanotechnologies. The development of microdischarges as light sources moved to the forefront mainly through the efforts to employ alternatives to mercury-based light sources. Additionally, some specified requirements, such as the need to produce thin large-area homogeneous sources, brought the focus of the nonequilibrium plasma community to the studies of the effects when one downsizes discharges, as well as possibilities to relate standard plasma (centimeter size and pressure close to 1 Torr) to microdimensions and pressure close to atmospheric. Table 3.27 tabulates a number of high-pressure microdischarges.

A large number of different configurations of micrometer-size discharges have been studied, but, generally, for the purpose of lighting, large-area flat panels comprised of arrays of microcavities proved to be the most promising. As far as the specific geometry of the individual elements of an array is concerned, the hollow-cathode-like structure is accepted as the most favorable and easiest to use in applications. There are several reasons for this. Discharges with hollow cathodes are the most stable, regardless of dimensions. In addition, due to recent advances in processing of materials, fabrication of nano- to micrometer size holes in different materials became easier and cost-effective, compared, for example, to parallel-plate discharges, which require careful treatment of cathode surfaces and precise alignment of electrodes.

The main aspect that brought microdischarges on the scene for application was the possibility to produce nonequilibrium plasma at atmospheric pressure. Most of the current plasma applications require nonequilibrium conditions. However, nonequilibrium plasma is produced easily only at a low pressure that requires high-cost vacuum technology, which is not appropriate for mass production. In standard centimeter-size discharges at atmospheric pressure, the breakdown results in a very rapid growth of ionization, where strong coupling between electrons and ions easily leads to the formation of a thermal plasma. There are several techniques for producing nonequilibrium plasma at atmospheric pressure or generally at a high pressure; by interrupting the discharge temporally or by applying inhomogeneous electric fields; by using a dielectric barrier; and by employing small electrode gaps, in the order of a micrometer.

Table 3.27. High-Pressure Microdischarges

UV Emitter	Peak Wavelength (nm)	Source Gas	Pressure (Torr)	DC Voltage (V), Otherwise Stated	Current (mA)		Efficiency (%)	
					DC	ms Pulses	DC/AC/ms Pulses	ns Pulses
HE^*_2	75	He	≤760	200	1–5	—	—	—
NE^*_2	85	Ne	≤1100	295–330	0.0005–0.04	—	—	—
AR^*_2		Ar	≤740	200–400	0.5–1.0	30	<1	—
Xe^*_2	172	Xe	≤1100	200	1–10	—	6	3–5
I^*_2	342	Kr/I_2	15/1	220	1–8	<70	6–9	20
ArF*	193	He(94%)+ Ar(5%) +F_2 (1%)	≤400	DC	—	—	2	—
XeO*	238	Xe/O_2 or $Ne/Xe/O_2$	≤300	500	1–8	—	—	—
XeI*	253.2	Xe/I_2	1–50/1	300–700	0.1–4.0	—	—	—
NeD*	285	Ne/D_2	500 Torr (0.5–5% D_2)	AC 20 kHz	—	—	—	—
XeCl*	308	Ne+Xe(1.5%)+HCl (0.06%)	≤1100	190	1–8	<200	3	—
ArD*	310–315	Ar/D_2	500 Torr (0.5–1%)	AC 20 kHz, 170–220 V rms	—	—	$<10^{-3}$	—
H*	121.6 (Lyman-α)	Ne/H_2	740/1–3	200–400	0.5–1.0	—	0.2	—
O*	130.5	At+O_2 (<1%)	≤1100	300	1–10	—	<1	—

Source: Karl H. Schoenbach, (2012), Sources of Ultraviolet Radiation, *IEEE Journal of Quantum Electronics*, 48:6 [10].

Generally, it is difficult to establish a good understanding of the operating conditions in microdischarges. Perhaps the best approach would be to start from low-pressure studies and extend them to high pressure and small dimensions by employing scaling laws. Discharges should scale according to several parameters: E/N—electric field to gas density ratio, which is proportional to energy gain between collisions; pd—proportional to the number of collisions; jd^2—current density times the square of the gap, describing space-charge effects; and, ω/N—frequency normalized by gas density, for high-frequency discharges.

Most of the applications that are currently being developed operate in hollow cathode geometry, in normal or abnormal glow regime. Recent studies have shown that these discharges do not appear to resemble low-pressure hollow cathode discharges and it was found that under the standard experimental conditions the hollow-cathode effect is not significant due to a small electron mean-free path. Such results raise the issue of the scaling of discharge properties as a function of pressure (p) and characteristic dimension (d). Since hollow cathode discharges are complex and have complex characteristics with a number of different modes, it is better to start from parallel plane microdischarges and follow the scaling over several orders of magnitude of the characteristic dimension d. The disadvantage of this geometry is that it is more difficult to achieve stable operation. On the other hand, the discharge behavior in simple geometry is easier to understand, it is easier to follow basic processes and their kinetics in gas breakdown and discharge maintenance, and it is certainly the easiest to model.

Thus far, the studies of parallel-plate microdischarges have shown that, at gaps above 5 μm, standard scaling works as long as the parameters are properly determined and well defined. The two most critical issues in determining adequate scaling parameters, which are commonly disregarded, are the actual path of the discharge and the effective area of the discharge.

Due to a tendency for the discharge to operate at the lowest possible voltage, at high pd-s it will operate at the shortest possible path between electrodes, while at low pd-s the discharge will establish along the longest possible path. Under conditions, where the mean-free path of electrons is in the order of micrometers, the discharge will tend to operate, not only between electrodes but also from the sides and back surfaces, unless that path is blocked. Therefore, effectively, parameter d does not have to be the actual gap between electrodes. This becomes even more important in complex geometries, such as those with hollow cathodes.

Another issue is also associated to the small mean-free path in microdimensions. Namely, proper determination of current density (i.e., scaling parameter jd^2) requires proper determination of effective discharge diameter. An electrode diameter is commonly used as the diameter of the discharge. However, as the discharge diameter depends on the diffusion length, which in turn depends on the electron mean-free path, it is obvious that at high pressure, the discharge will be quite constricted. Therefore, it is essential to use the actual diameter of the discharge in order to produce the scaling parameter jd^2. Again, in complex geometry, where it is hard enough to determine current density at a low pressure, this problem becomes even more difficult.

The analysis of volt-ampere characteristics is important to establish the regime of the discharge and also the breakdown voltage. There is an essential difference between Townsend (low-current diffuse), normal glow (constricted), and abnormal glow (high-current diffuse). In glow discharge, there is a cathode fall and possible constriction, which indicates importance of space charge effects. As normal glow has the lowest voltage of all the regimes, it is the most stable and usually the point when a transition, made from normal to abnormal glow, is made the most stable operating point. The large density of space charge leads to a complex field dependence where the mean energy, current, and other properties of electrons are not easily predicted and the range of conditions is limited. On the other hand, in Townsend discharge where the field is uniform and space charge does not cause deviations of the field profile, one can operate at different adjustable E/N making the discharge more efficient for excitation. Thus, it is favorable to use Townsend discharges for light sources as one could choose conditions where high-energy excited levels are excited efficiently and thus one can improve the efficiency of the light source. On the other hand, due to limited and very low current one cannot produce a large number of photons so the output is limited. With application of microdischarges due to jd^2 scaling, it is possible in microsize discharges to achieve very large current densities and consequently photon emission for the same jd^2. However, thus far there are no cases where microplasma light sources have operated in the Townsend regime.

Microcavity plasma has been applied for flat panel light sources that have a large variation of sizes, techniques to achieve microdischarges, and possible applications. In addition, similar sources may be used for the production of uniform glow discharges at atmospheric pressure, achieving power densities from tens to hundreds of kWcm^{-3}. Usually, such sources are applied for background lighting applications in both commercial and residential conditions. In production of these light sources, Al/Al$_2$O$_3$ are used. Thicknesses smaller than 1 mm have been achieved.

The microcavities that are used can be made in a number of ways. For example, a diamond may be used with Al foil (having characteristic dimension (D) of 250–1600 µm and encapsulated with 10–20 µm of nanoporous aluminum oxide (Al$_2$O$_3$). Another interesting way to manufacture microcavities is to use twisted wire bundles where in the small gaps between the wires a discharge forms with appropriate insulation of wires and applied power.

Arrays with active areas of more than 200 cm^2 have been reported for operation in the rare gases. Luminance of the order of 1700 cdm^{-2} has been achieved. For a Ne/20%Xe gas mixture illuminating a commercial green phosphor, a non-optimized luminous efficacy of 10.5 lm/W^{-1} has been obtained.

Future developments of the microdischarge-based light sources will be in the direction of improved or new microcavity geometries and technologies for their manufacture, in optimization of the efficiency and quality of light, and, finally, in developing applications with a wider range of applications.

Open microcavity sources may be used for other technologies including biomedical for large-area treatment at the atmospheric pressure of wounds and for sterilization. Further improvements are required for the luminosity and efficacy of the sources. One way to achieve both would be to use the Townsend regime, if possible, but a lot of research is required in this direction before practical success. On the other hand, much may be learned from standard-size discharges and further optimization or at least a better understanding of light source microdischarges may be achieved and should be aimed for.

Thin panel light sources have yet to achieve the efficacy and luminosity of their larger size counterparts. Research in scaling and fundamental discharge properties should lead to further improvements in this respect and may be the basis for developing competitive microdischarge-based light sources. In a way, a flat panel plasma TV may be regarded as one such source. Moreover, such discharges may be used as nonequilibrium plasma reactors for a wide range of applications, from the physics of materials to biomedicine.

References

1. Payling, R. and Larkins, P.L. (2000). *Optical Emission Lines of the Elements*. New York: John Wiley & Sons.
2. Jinno, M., Motomura, H., and Aono, M. (2005). Pulsed discharge mercury-free xenon fluorescent lamps with multi-pairs of electrodes. *Journal of Light & Visual Environment* 29 (3).
3. Kitsinelis, S. et al. (2004). Relative enhancement of near-UV emission from a pulsed low-pressure mercury discharge lamp, using a rare gas mixture. *J. Phys. D: Appl. Phys.* 37: 1630–1638.
4. Kitsinelis, S. et al. (2005). Medium pressure mercury discharges for use as an intense white light source. *J. Phys. D: Appl. Phys.* 38: 3208–3216.
5. Zissis, G. and Kitsinelis, S. (2009). State of art on the science and technology of electrical light sources: From the past to the future. *J. Phys. D: Appl. Phys.* 42: 173001 (16 pp.).
6. Hilbig, R. et al. (2004). Molecular Discharges as Light Sources. Proceedings of the 10th International Symposium—Science and Technology of Light Sources, Toulouse, France, p. 75.
7. Kitsinelis, S. et al. (2009). A strategy towards the next generation of low pressure discharge lamps: Lighting after mercury. *Journal of Physics D: Applied Physics* 42.
8. Huber, K.P. and Herzberg, G. (1979). *Molecular Spectra and Molecular Structure IV Constants of Diatomic Molecules*. New York: Van Nostrand Reinhold Company.
9. Pearse, R.W.B. and Gaydon, A.G. (1950). *The Identification of Molecular Spectra*, 2nd Edition. London: Chapman & Hall.
10. Schoenbach, Karl H. (2012). Sources of ultraviolet radiation. *IEEE Journal of Quantum Electronics* 48: 6.

④ Solid-State Light Sources

A ll artificial light sources on the market are based on three technologies, two of which (incandescence- and plasma-based sources, otherwise known as *discharges*) have reached a plateau with regard to efficacy but nevertheless there is still ongoing research and fascinating developments. The third technology, although its development and marketing started much later than these other two (and perhaps for that very reason), is now showing such fast progress that not only can it compete with the other technologies, but in some cases it is the preference.

This chapter is an introduction to this technology of solid-state light sources, known as *light-emitting diodes* (LEDs), which have attracted the interest of many professionals. The chapter covers a number of topics, answering the most frequently asked questions. Some topics include semiconductor and diode technology, on which LEDs are based, ways to create different colors and white light, modes of operation, thermal management, applications, and comparisons with the other two light source technologies. The chapter also addresses the wide range of professionals for whom light, its sources, and, in particular, the solid-state lamps are part of their work.

Solid-state lamps, that is, light-emitting diodes, are considered by many to be the future of lighting. Indeed, they have evolved to such an extent that they demonstrate several advantages over the other technologies and are already dominating some applications. Of course, there are still some limiting factors for their further development such as the issue of thermal management. A large number of scientists are working intensively on the technology of solid-state lamps and, more particularly, are focused on understanding how to create light through crystals, the reliability and performance of the materials in order to reduce production costs, the development of phosphor powders for conversion of radiation with good quantum efficiency, the geometry and materials of various parts for better extraction of photons, LED's sensitivity to temperature and humidity, and finally on improving the control electronics of the large number of units needed to produce high luminous flux and color reliability.

There are many who parallel the seemingly inevitable future dominance of LEDs with the case of the transistor where the hitherto dominant technology of thermionic valve/tube (glass-metal-vacuum) was replaced by a solid-state technology.

Finally, it is important not to forget that a similar but at the same time different technology of solid-state lighting is based on organic (OLED) and polymer (POLED) compounds. This technology may prove even more important in the coming decades if the materials used become cheaper and more flexible.

4.1   Light-Emitting Diodes

The technology of solid-state lighting was the last of the three to penetrate the market and is based on the effect of electroluminescence.

The term *electroluminescent* refers to light emission from a solid body when an electric current flows through it or when it is placed in an electric field, and its effect is different from incandescence. The first efforts to create light in this way focused on the use of phosphorescent powder such as ZnS (enriched with copper or manganese) in a powder or thin film form for use as a backlight for liquid crystal displays. These light sources consume little power but require high voltages (>200 V) while their efficiency is low. These efforts and the development of semiconductor technology that gave birth to the solid-state diode led to a new generation of solid-state light sources.

The controlled addition (doping) of small quantities of certain materials in a semiconductor's crystal structure such as silicon without damaging the structure (high quality crystals must be used avoiding oxygen and hydrogen) gives the semiconductor some extra properties and depending on the materials used there are two different cases. In one case, if the additional material consists of atoms with a number of valence electrons larger than that of the crystal atoms, we call the semiconductor *n-type* and it has a surplus of electrons in the crystal structure (for example, adding a small amount of phosphorus or arsenic in silicon). In the other case, where the added materials consist of atoms with a smaller number of valence electrons (adding a small amount of boron in silicon or gallium), the semiconductor has a surplus of positive charges, otherwise known as *electron holes*, and such a semiconductor is called a *p-type* (see Figure 4.1).

The connection that an n-type with a p-type semiconductor creates between them is called a *p-n junction*, which functions as a diode and allows the flow of electricity in one direction only, from anode (p-type) to cathode (n-type) as shown in Figures 4.2 and 4.3.

During the flow of electricity through such a solid-state diode, electrons are combined in the semiconductor junction with the positive holes and this combination puts the electrons in a lower energy state. The energy state difference can be released as electromagnetic radiation (not always, as it can also be lost as heat in the crystal) with a wavelength that depends on the materials of the semiconductor.

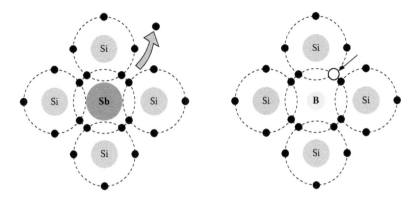

Figure 4.1. Creation of semiconductors with the addition of various elements to crystalline silicon (doping). The addition of antimony creates free electrons and the addition of boron creates electron holes.

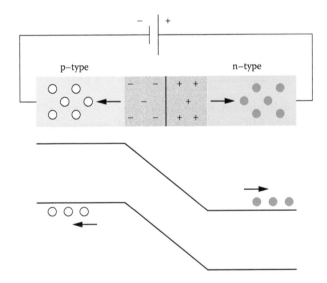

Figure 4.2. Creation of a diode by connecting n- and p-type semiconductors.

Such a light source is known as an *LED* (*light-emitting diode*). LEDs emit radiation of a narrow bandwidth (a range of a few tens of nanometers). The total charge in the crystal must be distributed as much as possible and the capsule should be optically transparent (epoxy). The flux is limited by the heat generated at the junction. Both the shell (epoxy) and the crystals begin to wear out after 125°C. Because of the low flux, LEDs currently have limited applications. Due to the materials used, total internal reflection takes place, which easily traps the radiation. The extraction techniques of photons are therefore an important aspect of this technology.

If the material is an organic compound, then we have an OLED (organic light-emitting diode) and in the case of a polymer compound the acronym used is POLED.

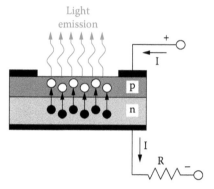

Figure 4.3. Radiation emission from the junction of a semiconductor during flow of electricity.

A brief historical review of the invention and development of this technology is provided below:

1907—H.J. Round, Marconi Labs, invents pale blue light from a SiC crystal.

1920s—Oleg Vladimirovich Losev develops first LED.

1955—Rubin Braunstein, Radio Corporation of America (RCA), reports the first infrared LED using GaAs, GaSb, InP, and SiGe.

1961—Bob Biard and Gary Pittman, Texas Instruments, file the first patent for an infrared LED using GaAs.

1962—Nick Holonyak, Jr., General Electric Company, builds the first red LED.

1968—Monsanto Corporation creates the first mass production of LEDs seeing a dramatic reduction in cost.

1972—Jacques Pankove, RCA, develops the first blue LED using GaN.

1972—M. George Craford develops the first yellow LED.

1976—T.P. Pearsall is responsible for the first use of an LED in telecommunications.

1990—High brightness red, orange, and yellow LEDs.

1993—Shuji Nakamura, Nichia Corporation, invents the first blue LED using InGaN (see also 2014 below).

1995—High brightness blue and green LEDs.

2000—White LED lamp with incandescent efficacy (20 lm/W).

2005—White LED lamp with fluorescent efficacy (70 lm/W).

2010—White LED lamps exceed 100 lm/W.

2014—The Nobel Prize in Physics awarded jointly to Isamu Akasaki, Hiroshi Amano, and Shuji Nakamura "for the invention of efficient blue light-emitting diodes which has enabled bright and energy-saving white light sources."

Figures 4.4 and 4.5 illustrate the anatomy of a modern LED unit, while Figure 4.6 shows a variety of shapes and forms in which LEDs can be found.

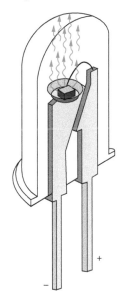

Figure 4.4. Light-emitting diode (LED) diagram.

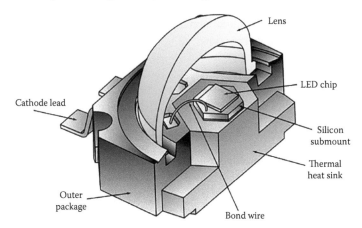

Figure 4.5. Solid-state lamp—light-emitting diode (LED).

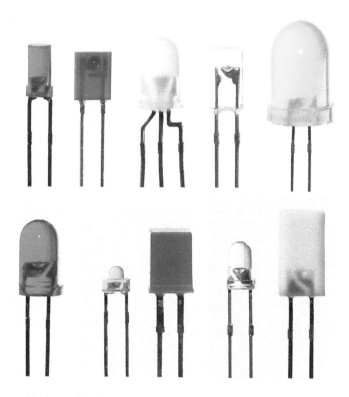

Figure 4.6. Variety of solid-state light sources (LEDs).

The following list is representative of compounds that with the appropriate addition of materials (doping) and the connection of the p-type and n-type semiconductors created, result in the emission of radiation after the flow of electricity. A general rule is that the energy difference increases (wavelength of emission decreases) with increasing aluminum (Al) concentration and decreases with increasing indium (In) concentration.

AlGaAs—Red and infrared

AlGaP—Green

AlGaInP—High brightness, orange-red, orange, yellow, and green

GaAsP—Red, orange-red, orange, and yellow

GaP—Red, yellow, and green

GaN—Green and blue

InGaN—Near ultraviolet, blue-green, and blue

SiC as substrate—Blue

Sapphire (Al$_2$O$_3$) as substrate—Blue

ZnSe—Blue

Diamond (C)—Ultraviolet

AlN, AlGaN—Ultraviolet

Figure 4.7 shows the emission spectra of three different LEDs in three different regions of the visible range. LEDs emit radiation of a relatively narrow bandwidth, which is shown in Figure 4.8.

The efficiency of LEDs is defined by several factors, such as:

- The electric efficiency that has to do with the number of charges in the material (>90% achieved).

- The internal quantum efficiency, which is the number of photons per number of electrons (this depends on the material and construction of layers—heat and re-absorption are the main problems).

- Extraction efficiency, which is the number of emitted photons per total number of photons (the geometry of the material and capsule plays an important role).

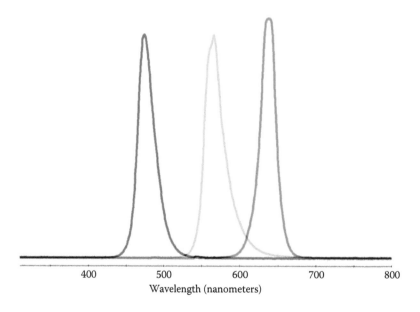

Figure 4.7. Emission spectra of three different LEDs in three different regions of the visible range.

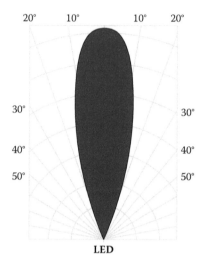

Figure 4.8. Light from LEDs is emitted in small solid angles.

- The spectral or optical efficiency, which is related to the eye sensitivity curve (this factor is not taken into account for an LED emitting at the limits of the curve).

4.2 Organic LEDs

If the material used is an organic compound, then it is known as an OLED (organic light-emitting diode). For the organic compound to function as a semi-conductor, it must have a large number of conjugated double bonds between the carbon atoms. The organic compound may be a molecule with a relatively small number of atoms, in crystalline form, or a polymer (POLED), which offers the advantage of flexibility (see Figures 4.9 and 4.10). For the time being, the organic LEDs offer lower luminous efficiency and lower average lifetimes than their inorganic cousins.

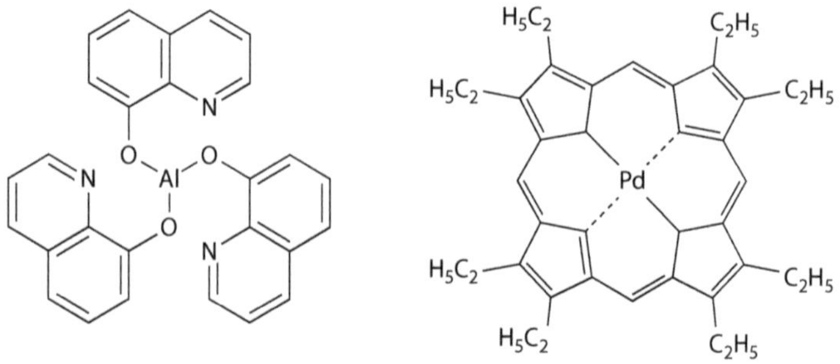

Figure 4.9. Chemical structures of organic molecules with double bonds, used for the development of OLEDs.

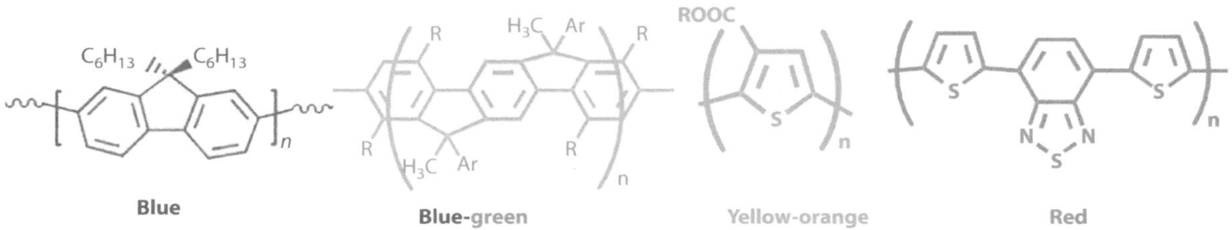

Figure 4.10. Fluorescent polymers. Tuning color by main chain structure. (Data from Mitschke, U. and Bauerle, P., 2000, The Electroluminescence of Organic Materials, *J. Mater. Chem.*, 10: 1471; Scherf, U. and List, E.J.W., 2002, Semiconducting Polyfluorenes—Towards Reliable Structure–Property Relationships, *Adv. Mater.*, 14: 477; Babudri, F., Farinola, G.M., and Naso, F., 2004, Synthesis of Conjugated Oligomers and Polymers: The Organometallic Way, *J. Mater. Chem.*, 14: 11.)

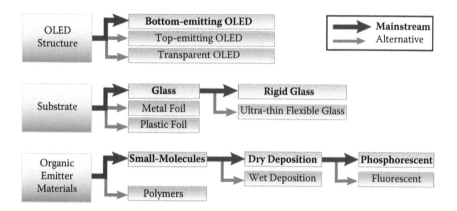

Figure 4.11. OLED structure paths.

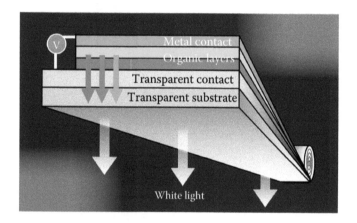

Figure 4.12. A schematic of an OLED structure.

Organic light-emitting diodes are steadily making their way into commercial devices such as cell phones and flat-screen displays. They are flexible, and use less power and less expensive materials than liquid crystal displays. Simplified descriptions of the OLED structures are illustrated in Figures 4.11 and 4.12.

A downside is that because the polymers react easily with oxygen and water, OLEDs are expensive to produce—they have to be created in high-vacuum chambers—and they need extra protective packaging layers to ensure that once they are integrated into display devices, they do not degrade when exposed to air or moisture.

Figure 4.13 shows some examples of commercial OLED sources.

OLEDs have the potential to grow into a very energy efficient light source. The ultimate potential is for the technology to reach efficiencies as high as 150 lumens per watt. A combination of these technologies can also lead to future light sources. One idea that has been proposed is a hybrid light-emitting diode, or HLED. This device would incorporate both organic and inorganic layers, combining the flexibility of an OLED with the stability of an inorganic light-emitting material.

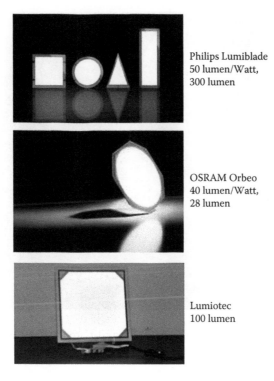

Philips Lumiblade
50 lumen/Watt,
300 lumen

OSRAM Orbeo
40 lumen/Watt,
28 lumen

Lumiotec
100 lumen

Figure 4.13. Commercial OLED sources.

4.3   LED White Light Emissions

White light can be created with the use of different color LEDs (red, green, blue, or yellow and blue, or four different colors) or by using a phosphor on a UV or blue LED (UV LED with a trichromatic powder or a blue LED with a yellow powder—YAG: Ce). Figure 4.14 presents the different methods.

With three or more primary LEDs (Figures 4.15 and 4.16), all colors can be created. Red LEDs are the most sensitive to temperature and therefore corrections need to be made as the LEDs heat up. Moreover, the light intensity and angle of incidence of each LED must match and mix appropriately in order to correctly

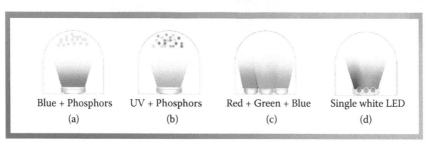

Figure 4.14. Ways to create white light.

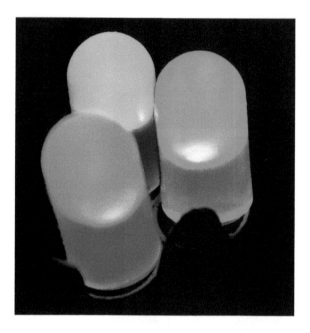

Figure 4.15. An example of a combination of different color LEDs for the creation of white or dynamic lighting.

Figure 4.16. Another combination of different color LEDs.

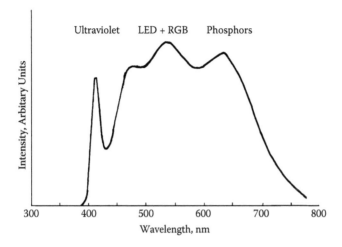

Figure 4.17. A UV LED can also be used with a trichromatic phosphor in order to produce white light.

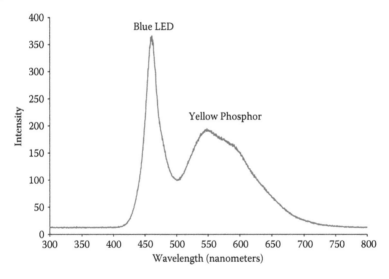

Figure 4.18. Blue LED emission spectrum with phosphor that converts part of the blue light into other colors/wavelengths (mainly yellow).

create the white light. The combination of blue and yellow light also gives the impression of white light since the yellow light stimulates the sensors of the eye that are sensitive to red and green but the resulting white light will be low on the color rendering index.

The other method for creating white light without using more than one LED is to convert ultraviolet or blue light into different colors by using a phosphor (Figures 4.17 and 4.18). The use of phosphor is less efficient due to Stokes' losses and other losses on the powder but it is an easier and cheaper way to create white light while the color rendering is usually better due to the larger spectral range of the powder. Another disadvantage of using a powder is the issue of distribution of light. The light emission angle from the crystal is different than that from the powder hence mixing of the different light wavelengths is not very good.

For blue to yellow conversion, the ratio of blue to yellow can be controlled through the quantity of the powder used, allowing control of the color temperature of the source. However, this method gives a source with a low color rendering index as there is a deficiency in red emissions, something that can be an issue in some applications such as general lighting but not in other applications such as signage. The light can have a color temperature of up to about 5500 K but with the addition of another powder that emits in the red part of the spectrum one can also create a warm white light temperature of 3200 K and better color rendering, at the cost of reducing the source efficiency.

The use of powder on a blue LED is the most economical way to create white light, and there are even proposals to use a blue LED (InGaN) with a green powder to replace the low-efficiency green LED or a blue LED with a red powder to replace the temperature sensitive red (AlInGaP) LED.

The use of phosphors on a UV LED can give white light of a significantly higher color rendering index but at the expense of efficiency (mainly due to Stokes' losses) in just the way a fluorescent mercury lamp operates. The different powders must be coated in such a way that there is no absorption of each other's emitted light. The powders used in fluorescent lamps are not appropriate as they are stimulated by the mercury emission lines at 185 and 254 nm while the UV LEDs emit at 360–460 nm.

With three LEDs there is better control over color (dynamic lighting) while the use of phosphors gives stability and a better mix. There are, of course, products that use two or three crystals in the same LED with appropriate wiring in order to create different colors and have better color mixing and control without the need for phosphors. The technology for using many crystals in the same LED, however, raises the cost due to the separate control of each diode, which requires more gear.

Whether one uses UV or blue LEDs with appropriate phosphors or uses suitable semiconductors, a variety of colors and accents of white can be produced today according to market demands.

To use three LEDs (each primary color) to create white light means that they have to be controlled during operation as they wear out differently and show different sensitivities to heat. The appropriate electronic and optical components can provide this control. When using phosphors one cannot control or make corrections and the increase in temperature shifts the emission wavelengths of blue LEDs.

Table 4.1 lists the advantages and disadvantages offered by the different ways to create white light with the use of light-emitting diodes.

A third method for producing white light without the use of fluorescent powders is through the combination of radiation simultaneously produced by the semiconductor and its substrate (blue radiation from ZnSe and yellow light from the ZnSe substrate). The absence of fluorescent powder means higher efficacy.

Table 4.1. Comparisons of Different Methods for Creating White Light

	Advantages	Disadvantages
Mixing Different Color LEDs	• Dynamic lighting • Ability to create millions of colors • Better efficacy • Control of component colors	• Different colors have different sensitivities to heat/no stability • Complex electronic gear • Not good color mixing
Blue LED with Phosphor	• Good efficacy • Good color rendering index • Wide color temperature range • Better stability	• Not good color mixing at certain angles • No control or regulation of different colors
Ultraviolet LED with Phosphor	• Good color mix • Wide color temperature range • Good color rendering index • Better stability	• Poor efficacy • Low power • Must manage UV light • No control or regulation of different colors

Finally, a method that is developing rapidly is the use of quantum dots—nano-crystalline semiconductor materials with dimensions equal to a few dozen atoms that emit light (fluoresce) with high efficiency under electrical or optical stimulation. The wavelength of the radiation can be controlled by altering the size of the nanocrystals and this method is in experimental stages (see Figure 4.19 for related spectra).

As LEDs emit narrow bands of light, the concept of CRI must be reconsidered taking into account saturated colors as shown in Figure 4.20.

Due to the low levels of emitted light from LEDs, a lamp must make use of multiple LEDs in order to be functional as a light source for general lighting. Many LEDs are needed for most applications as shown in Figure 4.21.

Most LEDs on the market operate at low power, usually less than 1 watt, but some products operate at powers as high as 7 watts with an efficacy similar to compact fluorescent lamps. For low-power LEDs, the efficacy exceeds the 100 lm/W mark. Generally, LEDs emit from 1 lm to several tens of lm and depending on their flux, LEDs are categorized as follows:

- Indicators with <5 lm

- Standard with 5–50 lm

- High brightness (HB) with 50–250 lm

- Ultra-high brightness (UHB) with >250 lm

If LEDs reach 250 lm/W in the next two decades, they could replace all fluorescent lamps that are currently limited to 50–120 lm/W.

A general rule that has been stated and so far seems to hold is the doubling of light output from each LED every 24 months in the last 40 years (similar to

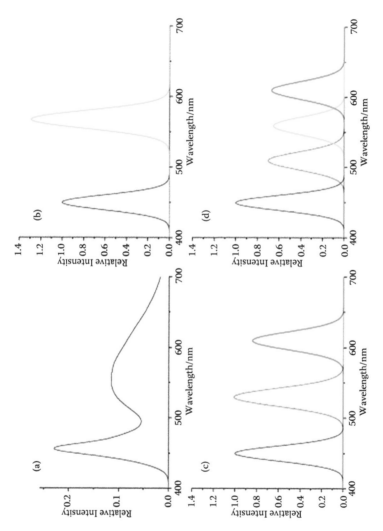

Figure 4.19. Various emission spectra of white LEDs. (a) Blue LED with phosphor while the other three spectra (b), (c), and (d), belong to dichromatic, trichromatic, and tetrachromatic quantum dot LEDs, respectively.

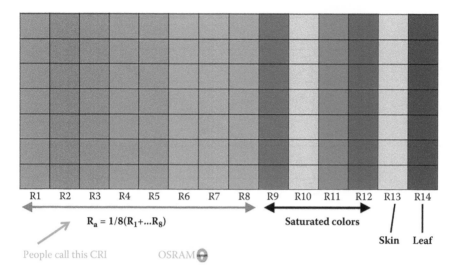

Figure 4.20. Color rendering index reconsidered with the coming of LEDs to dominance.

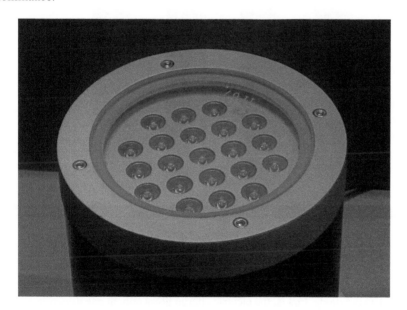

Figure 4.21. An LED lamp comprised of many LED units.

Moore's law) and the cost for LEDs is halved almost every decade. Figures 4.22 and 4.23 report some trends in cost and performance. This pattern was first observed and described by Roland Haitz. In most cases, LEDs operate with less than 100 mA and when the current exceeds that value, they are referred to as *power LEDs*. For high-power LEDs, a silicon gel is used instead of an epoxy coated by a polymer, and an appropriate heat sink is used for proper heat management.

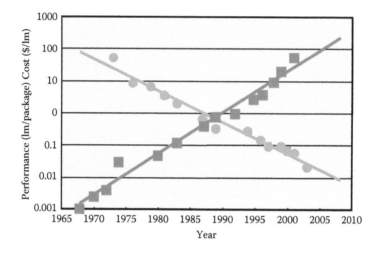

Figure 4.22. The cost of the LED is continuing to decrease while LED performance improves.

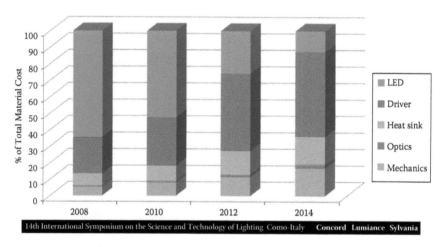

Figure 4.23. The LED part of the system cost has decreased significantly over time.

4.4   LED Operation

The LED is a diode, that is, the electric charge flows only in one direction and not the opposite. Once the LED starts to operate normally, the current is linearly proportional to the voltage, namely, the relationship between the voltage and current for LEDs is positive as seen in Figure 4.24.

$$Vf = Vo + Rs\ If$$

Vo is the initial voltage that must be applied before the charges start to flow. The value depends on the material and the energy difference between the states

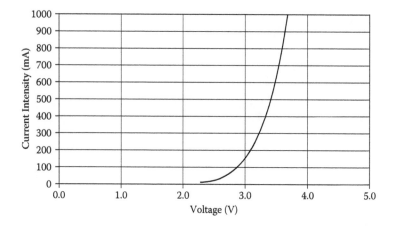

Figure 4.24. Voltage–current relationship in the starting and operation of an LED.

(band gap). In general, the voltage is influenced by the temperature of the diode. The resistance also depends on the material and is quite low.

During production, there are shifts in Vo and the resistance so the categorization of the final products according to their electrical characteristics (binning) is necessary as the light output is proportional to the intensity of the current and therefore the voltage.

The two most important factors in limiting light emission are heat and current density. At large currents, the materials wear out mainly due to high temperatures but also due to the leaking of charges to other layers outside the active crystals (especially when the layer of the active crystal is thin). The maximum values depend on the material but, in general, more than 120 A/cm² is prohibitive for all cases. High temperature not only damages the crystal but also causes shifts in the emitted spectrum. For example, yellow can be converted to red at high temperatures just before the destruction of the crystal and the diode junction. This change corresponds to about 0.1 nm per degree Celsius, something that is a great disadvantage if color reliability is important.

LEDs generate heat during operation (not in emission) and the materials are very sensitive to it. Thus, the thermal design is an important part of the design of these light sources.

An LED is a low-power light source so LED lamps cannot be connected directly to the mains voltage as they need a current and voltage controller to keep them at low values and also to run several LEDs at once (as shown in Figures 4.25 and 4.26). In other words, LED lamps need the equivalent of a ballast, which we call a *driver*.

The driver can be a constant voltage driver or a constant current driver.

The constant voltage driver is not as reliable as there is a differentiation of the required voltage during production and because the resistance is small, variations in voltage result in significant variations in intensity.

The constant current driver is preferable as the intensity is set to the normal operating intensity of the LEDs and the voltage is adjusted.

Figure 4.25. An example of LEDs in series mainly used for decorative lighting.

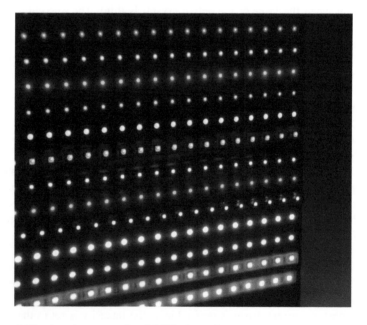

Figure 4.26. Another example of LEDs in series.

There is also a hybrid driver of constant voltage with a large resistor parallel to the LED. However, the power consumption by the resistor renders such drivers inefficient.

A driver can be used for many LEDs and how to connect them (series/parallel) is important. If we have a constant voltage driver then a resistor should be added so that we have a nominal current intensity while with constant current drivers the number of LEDs needs to be taken into account. In each case, the current intensity through each LED or series of LEDs must be estimated carefully.

The position of each LED in a system plays an important role in the functioning of the whole layout. If an LED fails, it can create an open circuit (e.g., in a series connection) or a closed circuit with different electrical characteristics (e.g., if all are parallel) such as higher current in one of the series leading to asymmetric production of light. A combination of series and parallel connections offers greater reliability because the current can find alternative paths without overloading some of the units.

Protection from large positive (normal flow) and negative (opposite polarity) voltage is provided by high voltage diodes in appropriate connections. LEDs withstand very short (<1 ms) and nonrecurring high current pulses (hundreds of mA).

Of course, there are intelligent LED control systems, which cannot only create a wide range of colors by combining multiple light sources and offer dynamic lighting but can also take into account variations in temperature or other electrical characteristics and make the appropriate corrections and changes. There are many communication protocols to control lighting systems and the choice depends on the application. It is important for an intelligent LED driver to be able to receive and analyze the following list of signals.

- Linear voltage control (0 to 10 V)

- Digital multiplex (DMX512)

- Digital addressable lighting interface (DALI)

- Power-line communication (PLC)

- Domotic standards: INSTEON, X10, Universal Powerline Bus (UPB) and ZigBee

4.5 Thermal Management of LEDs

All light sources convert electrical energy into heat and radiation emitted in varying ratios. Incandescent lamps emit mostly infrared radiation and a low percentage of visible light. Low- and high-pressure discharge lamps (fluorescent and metal halide, respectively) produce more visible light but also emit infrared and ultraviolet radiation as well as heat. LEDs do not emit infrared and undesirable ultraviolet radiation and besides the visible light produced, the remaining energy is converted into heat, which must be transferred from the crystal to the circuit

Table 4.2. Power/Energy Conversion for Different *White* Light Sources

	Incandescent	Fluorescent	Metal Halide	LEDs
Visible light	10%	20%	30%	15–25%
IR	70%	40%	15%	~0%
UV	0%	0%	20%	0%
Emitted energy	80%	60%	65%	15–25%
Heat	20%	40%	35%	75–85%
Total	100%	100%	100%	100%

(as the capsule surrounding the LED is not thermally conductive, the heat flows to the other direction) and from there to other parts of the unit until finally being transferred to the environment through the air. See Table 4.2 for a comparison of different technologies regarding energy conversion.

The removal of heat happens first with its flow from part to part and then from the surfaces to the environment. This means that many factors play a role in how efficiently and quickly the heat is removed. Some of these factors include:

- The materials, which are made of different parts (preferred materials are those which are thermally conductive, i.e., low thermal resistance such as metals, but new polymers also exhibit good conductivity).

- Their connections (there should be a good contact between each part and gaps do not help because air is not a good conductor of heat).

- Their total surface area (large areas with as less volume as possible).

Heat production (and the temperature increase at the diode junction which accompanies it) is the major limiting factor and the biggest obstacle to developing LEDs with higher power and brightness. Therefore, the issue of thermal management is currently perhaps the most important problem that scientists and technologists have focused on and their efforts are directed toward finding and using material of high thermal conductivity to reduce as much as possible the thermal resistance, $R\theta$, of the system so that heat is removed as quickly and easily as possible.

The thermal resistance, $R\theta$, of an LED is defined as the ratio of the temperature difference between the junction and the environment (ambient temperature) over the consumed power due to the current that flows through the LED.

$$R\theta = (\Delta T \text{ junction} - \text{ambient})/P$$

where
$\Delta T = T \text{ junction} - T \text{ ambient}$
$P = \text{current intensity (I)} * \text{voltage (V)}$

The total thermal resistance must, of course, include the entire system from the junction to the surfaces that are in contact with the surrounding air. Each individual part of an LED is characterized by a different thermal resistance (whose values

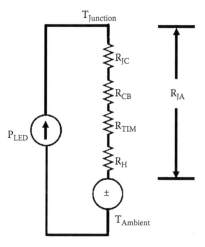

Figure 4.27. Thermal model of an LED *circuit*.

depend on the geometry, material, and surface area of each piece) and according to these we can define the thermal resistance of the crystal, which is given by the manufacturers (the smaller the value, the easier the transfer of heat). Figure 4.27 shows a thermal model for an LED circuit, which is an analogy to an electrical circuit.

By knowing the thermal resistance provided by the manufacturer between the junction and the material on which the crystal rests, and by measuring the temperature differences between the other parts, using infrared detectors or thermocouples, the temperature at the junction for different current intensities and power values can be calculated.

The thermal resistance with units °C/W, is deduced from the thermal conductivity (with unit W/mm), the length of the heat conductor, and its cross-section. This practical parameter allows us to calculate various temperatures at different parts of the system when the consumed power is known. The model used is that of an electrical circuit where the parallelisms are:

Heat Q (watts) ~ Current intensity

Thermal resistance $R\theta$ (°C/W) ~ Electrical resistance

Temperature difference ΔT (°C) ~ Voltage

The equivalent to Ohm's law is $\Delta T = Q \times R\theta$

4.5.1   What Defines the Junction Temperature?

Light and heat are produced at the junction of the diode, which has small dimensions so the heat production per unit surface area is very large. The temperature of the junction cannot be measured directly but it can be calculated by measuring the temperature of another part and taking into account the thermal resistances of all materials.

There are three factors that determine the junction temperature of an LED: the intensity of the operating current, the thermal resistance of the system, and the ambient temperature. Generally, the greater the intensity of the current, the more heat is produced in the crystal. The heat must be removed in order to maintain the flux, the lifetime, and the color. The amount of heat that can be removed depends on the temperature and the thermal resistances of the materials that make up the whole LED.

The products on the market have a maximum temperature at which they must operate which is around 120°C. The efficacy and lifetime, however, begin to decline well before that temperature limit. Very few power LEDs have the appropriate initial design that allows them to function at maximum power without using a secondary cooling system. Temperature increase without proper control and stabilization is certainly the main reason for early destruction of LEDs. Although 120°C is given as the maximum operating limit, a more realistic limit is that of 80°C as one must take into account that fluctuations of the ambient temperature can be of the order of 25°C or higher.

Manufacturers classify LEDs according to the luminous flux and color under pulsed current (25 millisecond pulses) keeping the junction temperature constant at 25°C. But under normal operating conditions, the junction temperature is at least 60°C, so the flux and the color will be different than that of the manufacturer's specifications. The worst is the thermal management of an LED, then the larger will be the efficacy losses.

The rise in junction temperature has various consequences such as reductions in efficiency, life expectancy, and voltage value as well as shifts in the radiation wavelengths. The latter especially affects the operation of mostly white LEDs, causing changes in color temperature (see Figure 4.28 and Table 4.3).

In general, temperature affects each color to a different extent, with red and yellow LEDs being the most sensitive to heat and blue being the least sensitive.

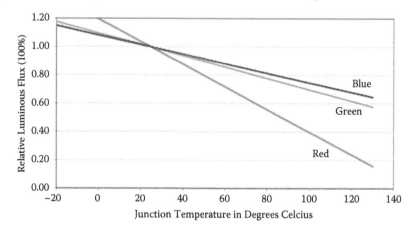

Figure 4.28. The effect of temperature on lifetime and flux for different colored LEDs presented for known commercial products.

Table 4.3. As the Temperature Rises
There Are Different Wavelength Shifts
for Different Color LEDs

Color	K (nm/°C)
Yellow	0.09
Red	0.03
Blue	0.04
Green	0.04

These different responses of each color to temperature can lead to changes and instabilities of the white light produced by RGB systems if during operation the junction temperature, Tj, is different from that specified by the manufacturer. But even when phosphors are used, the shifts are significant because the powders are sensitive to specific wavelengths.

4.5.2   Overheating Avoidance

There are two ways to control the junction temperature in order to avoid a malfunction or premature destruction of an LED. One is to decrease the intensity of the current so the LED is operated at a lower power and the other is to create a proper design so that the overall thermal resistance of the LED is minimized. The second, although preferable, is not as simple and requires taking into account several parameters such as the total surface area of the materials in contact with the surrounding air. The greater the contact area, the lower the thermal resistance.

A combination of both ways is common practice but proper calculations must be done in each different case. Since the flux decreases with both increasing temperature and decreasing current and thus the power, one must work out whether (taking into account the thermal resistance given by the manufacturer) a decrease in power, which means operation with lower brightness, ultimately pays off. The voltage also depends on the junction temperature and it is reduced as the temperature increases. When dealing with a single LED or several in a series then the current can be controlled, otherwise there is a risk of having uneven distribution of the current intensity due to temperature variations and ultimately destruction of the system.

The circuit connected to the LED may include some type of heat sink, such as the MCPCB LEDs (metal-core printed circuit board) but higher power commercial LEDs use extra heat sinks characterized by their large size relative to the LED and their large surface area. A typical geometry of such extra large heat sinks includes the use of fins that dispose heat more efficiently. Such a heat sink is incorporated in luminaires or can be itself the luminaire. Finally, some LEDs have the circuit at a distance so that the heat generated by the circuit does not contribute to the increase of the junction temperature. See Figures 4.29 and 4.30, which display LED lamps with incorporated heat sinks.

In addition to electrical, mechanical, and optical data, a compact model for Flotherm software is now permanently available at: www.osramos.com/thermal-files,

Figure 4.29. An LED lamp with heat sinks incorporated in the design.

Figure 4.30. Another example of an LED lamp with heat sinks.

for calculating thermal behavior. Customers can find all the documents and the latest data needed for calculating thermal variables for different designs without building costly prototypes or carrying out time-consuming measurements. The data is available for standard high-power LEDs in the visible range, particularly for the Dragon family, the Power TopLED range, the Advanced Power TopLED range, and some Ostar variations.

The compact model available on the Web site is a simplified thermal geometry model that can be integrated in Flotherm software and used for customer-specific calculations. It is suitable, for example, for calculating the temperature distribution in a planned system.

4.6   Dimming/Controlling Brightness

An LED can be dimmed by controlling the current and this can be done in two ways:

- Increasing/decreasing the intensity (DC dimming)

- Pulse control

Controlling the DC current intensity has some disadvantages. As binning of LEDs takes place under the operation current by changing the intensity there is no longer reliability regarding common features between LEDs of the same category (proposed dimming to one-fourth of the intensity only). Low intensity also means big changes in parallel LEDs due to changes in voltage and, finally, changes in intensity can lead to differences in color.

On the other hand, with pulse control the maximum pulse current intensity value is set at the normal operating value and thus no changes in the characteristics while the mean or average intensity is defined by the frequency and duration of pulses. In this way, the LED works properly and a linear relationship between current intensity and luminous flux is assured. Controlling the duration of the pulses is an easy task because LEDs respond instantly ($<$ μs).

The dimming ratio is defined as the minimum mean intensity value over the maximum mean intensity of LED current. This percentage is determined by the shortest possible pulse that the driver can deliver and which at its maximum reaches the nominal current intensity of LED operation. The shorter pulse is, in turn, defined by the rise and fall times of the pulses.

This method makes dynamic lighting possible and it is imperative that the minimum frequency chosen should be one where the source is comfortable to the human eye.

Of course, with a combination of pulse control and DC current intensity checking, even smaller dimming ratios are possible.

4.7   General Characteristics of LEDs

A list summarizing the general characteristics of LEDs follows (see also Table 4.4).

- Their emission spectra are narrow band (a few tens of nanometers).

- They are characterized by their low brightness (flux) starting from 1 lm for conventional LEDs and reaching 120 lm for high-power LEDs.

Table 4.4. General Characteristics of LEDs

Efficacy lm/W	<130
Color temperature/K	Wide
Color rendering index	−90
Lifetime	50,000–100,000
Applications	Signage, remote control, fiber optic communication, decoration, advertising

- Their efficacy is increasing rapidly for market timescales and presently they exceed that of incandescent lamps and compare to discharge lamps such as low-pressure fluorescent and high-pressure metal halides.

- For high flux, many LED units are necessary.

- Wide range of colors with RGB mixing.

- They respond instantly without the switching frequency wearing them out. LEDs with phosphor have a slightly slower response time due to the powder fluorescence.

- Good brightness control and dimming ratio reaching 1/3 with current intensity decrease and 1/300 with pulse control (a few hundred Hz).

- Their end of life is characterized by a gradual decrease of the luminous flux and is not sudden.

- They are sensitive (and wear out if exposed) to heat, static electricity, and, in the case of blue and ultraviolet LEDs, to radiation. Generally though, they are characterized by their material strength in contrast to the fragile technology of glass and gas.

- Characterized by their small size, which means freedom in luminaire design. For large fluxes, however, the luminaire must also be large.

- Their average lifetime (70% of initial lm value) is 50,000 hours. The end of life depends on defects in the crystalline structure of the semiconductor or the fluorescent powder.

- Absence of harmful radiation in the ultraviolet and infrared part of the spectrum compared to other lamps, but sensitivity and reduced efficacy at elevated temperatures caused by the current flow. Almost all LEDs have an upper limit of 125°C.

- LEDs are still expensive not just because of their materials and manufacturing cost, but also due to their low luminous flux. Nevertheless, they offer a low-power consumption solution in applications that require low levels of light.

Figure 4.31 presents comparisons between LED and induction lamp technologies, while Figure 4.32 illustrates the comparisons between LED and OLED panels.

4.8 Applications

Advantages and disadvantages for different applications follow. The applications where LEDs have the advantage include:

- Those applications that require light of specific color. LEDs produce specific colors more efficiently than putting filters in incandescent lamps.

	Induction	LED
Source efficiency	Typical 90 to 130 Lm/W	Up to 130 Lm/W (theoretical > 280)
Power source	DC	DC
Lifetime (hours)	30,000 to 50,000	50,000 + (with proper system design)
CRI	Up to 95	80–90
CCT	2000 to 10,000	2000–6000
Instant light on	30s to full power - 60s for re-strike.	Yes
Dimmable	To 20% of full power	100%
Brightness	Up to 25,000 lm for a few mm	Typ. 130 lm for a mm^2 device driven at 1 W
Directionality	Yes (lambertian pattern)	Yes
Mercury	Yes (small amount)	No

Figure 4.31. Comparisons between LED and induction lamp technologies.

LED Panels

OLED Panels

Lumitronix LED Panel

System efficacy	= 72 lm/W (system)
Dimensions	= 300 × 300
Thickness	= 13 mm
Luminous flux	= 1350 lm
LT70	= 20 kh
Cost per lm	= 177 Eur/klm (system)
Source:	www.leds.de

LGchem OLED Panel *N6SD*30

Device efficacy	= 60 lm/W (panel)
Dimensions	= 320 × 320
Thickness	= 1 mm
Luminous flux	= 850 lm
LT70	= 40 kh
Cost per lm	= 200 USD/klm
Source:	www.LGchem.com

Philips

Figure 4.32. A comparison of commercial LED and OLED panels.

- Applications where long lamp lifetimes are required due to difficulty or high cost in replacements.

- Wherever small size light sources are required such as decorative lighting or small spaces (mobile phones, car interiors, etc.).

- Wherever instant start and dimming are necessary.

Applications where LEDs have disadvantages and must not be used widely yet include the following:

- In high temperature environments.
- Where high brightness is required.
- Where color stability is necessary.
- If accurate stability of color temperature and color rendering index is essential.
- If good knowledge of lifetime and lumen depreciation is needed.

The goal of LED companies is for this light source technology to dominate the following applications in the near future:

- TV and computer large screens
- Small projection systems
- Car headlights
- Interior lighting

Organic/polymer LEDs (OLEDs/POLEDs) will also play an important role in these applications but for now they are characterized by a short lifetime.

The different applications can be categorized into those where visual contact with the source is necessary (signage) and those where the reflected light is used (general lighting).

4.8.1   Signage: Visual Contact

In applications where high levels of brightness are not needed but the creation of optical signals of specific color is the aim, LEDs have already started dominating the market. Some of these applications include the following (LED characteristics that offer an advantage are in parentheses):

- Traffic lights (color, sturdy)
- Car taillights (style, size, electrically compatible) as shown in Figure 4.33
- Car interiors (size, mercury-free)
- Decorative lighting (size, dynamic lighting, dimming) as shown in Figure 4.34
- Monitors and screens (color mixing)
- Signs (long lifetime, size) for roads or shops (Figure 4.35)
- Mobile phones (size, low voltage)

Figure 4.33. The use of LEDs for automotive brake lights.

Figure 4.34. Decorative lighting with strips of red LEDs.

Figure 4.35. The signage industry has been reinvented with LEDs.

4.8.2   General Lighting: No Visual Contact with the Source

There are some applications in general lighting where LEDs are preferred for various reasons, which include the following:

- Dental treatments/bleaching (blue color, size, long lifetime, replacement of low efficiency halogen lamps)

- Flashlights (size, low voltage)

- Architectural lighting

- Machine sensors (various geometries, durability, sturdiness)

- Telecommunications (instant start)

LEDs have found a wide range of applications from building lighting to large panel displays. For LEDs to dominate in other markets the following characteristics must be improved or changed.

- Crystal materials and layer geometries so that the efficacy is increased (internal and extraction).

- The materials of all parts of the unit so that they can withstand a higher temperature thus higher power input in order to increase the brightness/luminous flux.

- Cost, which can be lowered with the help of the aforementioned changes.

- Better quality of white light with higher color rendering index and a wider range of color temperature.

- The production line so that there is greater reliability and no differentiation in efficiency.

4.9   Closing Remarks

The future of LEDs and perhaps of lighting, in general, will depend on the outcome of various research paths that scientists around the world are pursuing or must undertake. It is essential that a better understanding of the light generation mechanism be achieved as well as improvements in internal quantum efficiency. The LED manufacturing techniques must also improve so that consistency and better quality can lead to better marketing of LEDs. Besides the actual junction, the other parts of an LED are also crucial and must be researched and developed such as substrates, packaging, and lenses with proper thermal management. The ultimate goal is not only to increase the efficacy of each LED but also to achieve long lifetimes and tolerance to high temperatures. In addition, it is important to develop phosphors able of absorbing and converting photons efficiently and as it is the case already for fluorescents lamps, the development of *quantum-splitting phosphors* could be a breakthrough for LEDs.

The technology of OLEDs also has the potential to experience a rapid growth and market penetration as the great variety of organic luminescent materials could give a large number of emitting colors.

As with any other technology, high efficacies, consistency, long lifetimes, good color stability, uniformity over large surfaces, and relatively low manufacturing costs are the characteristics desired for a technology to succeed.

Standards

IESNA LM-79-08

ANSI C82.2 (efficacy)

IESNA LM80-08 (lumen depreciation)

ANSI C78.377A (CRI)

Test Standards

ANSI C82.77-2002 (PFC)

EN61000-3-2 (harmonics)

EN61000-3-3 (flicker and voltage variations)

Bibliography

Antoniadis, Homer. OSRAM. Overview of OLED Display Technology. http://www.ewh.ieee.org/soc/cpmt/presentations/cpmt0401a.pdf.

Bergh, A., Craford, G., Duggal, A., and Haitz, R. (2001). The promise and challenge of solid-state lighting. *Physics Today.* December 2001. http:// scitation. aip.org/journals/doc/PHTOAD-ft/vol_54/iss_12/42_1.shtml.

International Light Technologies. LED Technical Application. www.intl-lighttech.com/applications/led-lamps.

Khan, M.N. (2014). *Understanding LED Illumination.* Boca Raton, FL: CRC Press/Taylor & Francis.

Khanna, V.K. (2014). *Fundamentals of Solid-State Lighting: LEDs, OLEDs, and Their Applications in Illumination and Display.* Boca Raton, FL: CRC Press/Taylor & Francis.

Maxim Integrated. LEDs are still popular (and improving) after all these years. www.maxim-ic.com/appnotes.cfm/appnote_number/1883.

Rensselaer Research Lighting Center. LEDs. www.lrc.rpi.edu/researchareas/leds.asp.

Schubert, E.F. (2006). *Light Emitting Diodes,* 2nd Edition. Cambridge: Cambridge University Press, 2006.

Schubert, Fred E. Light-Emitting Diodes. www.ecse.rpi.edu/~schubert/Light-Emitting-Diodes-dot-org.

U.S. Department of Energy. National Energy Technology Laboratory. www.netl.doe.gov/home.

Žukauskas, A., Shur, M.S., and Gaska., R. (2002). *Introduction to Solid State Lighting.* New York: Wiley.

 5 Lasers

5.1 The Development of LASERs

A class of light sources which are not used for general lighting but for a wide range of other applications due to their unique characteristics is LASERs (Light Amplification by Stimulated Emission of Radiation). LASERs were developed out of the technology of MASERs (Microwave Amplification by Stimulated Emission of Radiation).

A concept for using lasers to create white light sources for general lighting is illustrated in Figure 5.1 but thus far it has not been deemed practical for development.

The emitted laser light is a spatially coherent (waves of the same frequency and phase), narrow low-divergence beam. The laser's beam of coherent light differentiates it from light sources such as lamps that emit *incoherent* light, of random phase varying with time and position. Although laser light is usually thought of as monochromatic, there are lasers that emit a broad spectrum of light (short pulses) or simultaneously at different wavelengths.

The invention of the laser can be dated back to 1958 with the publication of the scientific paper, "Infrared and Optical Masers," by Arthur L. Schawlow and Charles H. Townes, who were both working for Bell Labs. That paper launched a new scientific field and opened the door to a multibillion-dollar industry. The work of Schawlow and Townes, however, can be traced back to the 1940s and early 1950s and their interest in the field of microwave spectroscopy, which had emerged as a powerful tool for working out the characteristics of a wide variety of molecules (MASER technology was the first to be developed). Other important figures in the development of this technology are Gordon Gould, who first used the term *LASER* in his paper entitled, "The LASER, Light Amplification by Stimulated Emission of Radiation," and Alexander Prokhorov, who worked independently on this technology.

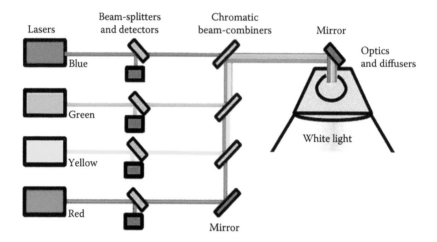

Figure 5.1. Combining different lasers to produce white light sources.

Lasers work by adding energy to atoms or molecules, so that there are more in a high-energy ("excited") state than in some lower-energy states; this is known as a *population inversion*. When this occurs, light waves passing through the material stimulate more radiation from the excited states than they lose by absorption due to atoms or molecules in the lower state. This "stimulated emission" was also the basis of masers.

Essentially, a laser consists of an amplifying or gain medium (this can be a solid, a liquid, or a gas) and a system that excites the amplifying medium (also called a *pumping system*). The medium is composed of atoms, molecules, ions, or electrons whose energy levels are used to increase the power of a light wave during its propagation. What must be achieved is a population inversion in the energy levels of the species followed by stimulated emission.

The pumping system creates the conditions for light amplification by supplying the necessary energy that will lead to population inversion. There are different kinds of pumping systems: optical (flash lamps, other lasers, etc.), electrical (discharge through the gas, electric current in semiconductors), or even chemical (reactions that lead to products with population inversion).

Theodore Maiman invented the world's first laser, known as the *ruby laser,* in 1960. A ruby crystal is composed of aluminum oxide doped with chromium atoms. In a ruby laser, a ruby crystal is formed into a cylinder. A fully reflecting mirror is placed on one end and a partially reflecting mirror on the other. A high-intensity lamp around the ruby cylinder provides the energy in the form of white light that triggers the laser action. The lamp flash excites electrons in the chromium atoms to a higher energy level. Upon returning to their normal state, the electrons emit their characteristic ruby-red light. Mirrors at each end reflect the photons back and forth stimulating other excited chromium atoms to produce more red light, continuing this process of stimulated emission and amplification. The photons finally leave through the partially transparent mirror at one end.

The ruby laser is still used, mainly as a light source for medical and cosmetic procedures, and also in high-speed photography and pulsed holography.

5.2   Output Types

The output of a laser may be continuous (known as *CW* for *continuous wave*) or pulsed. In the continuous mode, the population inversion required for lasing is continually maintained by a continuous pump source. In the pulsed mode of operation, one can achieve higher power outputs per pulse for the same average energy usage. In order to produce laser pulses, several techniques are used such as Q-switching, mode-locking, and gain-switching.

In a Q-switched laser, the population inversion is allowed to build up by making the cavity conditions unfavorable for lasing. Then, when the pump energy stored in the laser medium is at the desired level the system is adjusted releasing the pulse. This results in high-peak power as the average power of the laser is packed into a shorter time frame. The Q-switch may be a mechanical device such as a shutter, chopper wheel, or spinning mirror/prism placed inside the cavity, or it may be some form of modulator such as an acousto-optic device or an electro-optic device—a Pockels cell or Kerr cell. The reduction of losses (increase of Q) is triggered by an external event, typically an electrical signal, so the pulse repetition rate can therefore be externally controlled. In other cases, the Q-switch is a saturable absorber, a material whose transmission increases when the intensity of light exceeds some threshold. The material may be an ion-doped crystal like Cr:YAG, which is used for Q-switching of Nd:YAG lasers, a bleachable dye, or a passive semiconductor device. Initially, the loss of the absorber is high, but still low enough to permit some lasing once a large amount of energy is stored in the gain medium.

Mode-locking is a technique where a laser can be made to produce pulses of light of extremely short duration, on the order of picoseconds (10^{-12} s) or femtoseconds (10^{-15} s). The production of such short pulses has allowed the possibility to explore really fast physical and chemical events (such as chemical reactions), as well as the study of nonlinear effects in optics. The basis of the technique is to induce a fixed phase relationship between the modes of the laser's resonant cavity. The laser is then said to be *phase-locked* or *mode-locked*. Interference between these modes causes the laser light to be produced as a train of pulses. A pulse of such short temporal length has a spectrum which contains a wide range of wavelengths. Because of this, the laser medium must have a broad enough gain profile to amplify them all. An example of a suitable material is titanium-doped, artificially grown sapphire (Ti: sapphire).

Yet another method for achieving pulsed laser operation is to pump the laser medium with a pulsed pumping source, which can be a flash lamp or another laser. Such a technique is common in the case of dye lasers where the inverted population lifetime of a dye molecule is so short that a high-energy, fast pump is needed. Pulsed pumping is also required for lasers that disrupt the gain medium so much during the laser process that lasing has to cease for a short period of time.

These lasers, such as the excimer laser and the copper vapor laser, can never be operated in a CW mode.

The aforementioned techniques have led to ever decreasing pulse durations and ever increasing peak powers per pulse. In recent years, certain facilities around the world have produced extreme values such as the following:

> The National Ignition Facility (NIF) at the Lawrence Livermore National Laboratory in Livermore, California, recently achieved 150,000 joules in a single 10 nsec pulse. One application of such a laser is to direct this energy into a tiny pellet containing deuterium in an effort to induce nuclear fusion for the development of an abundant source of energy.

> At the Max Planck Institute for Quantum Optics in Garching, Germany, a pulse of just under 1 femtosecond, or 10^{-15} sec, was produced (this is less than the time photons take to complete a single oscillation cycle). In order to make such a short pulse, the final result must be a white burst containing all colors from the visible region well into the ultraviolet region of the spectrum. This is very different from typical lasers, which emit a single color, but such short pulses cannot be so monochromatic due to the uncertainty principle.

> The Lawrence Livermore National Laboratory produced pulses of over 1 PW, or petawatt (10^{15} W), a few years ago. Because laser light can be focused to a very small spot, the focused energy density reached the equivalent of 30 billion joules in a volume of 1 cubic centimeter, far larger than the energy density inside of stars.

> The Mid-Infrared Advanced Chemical Laser (MIRACL), at the High Energy Laser Systems Test Facility at White Sands Missile Range, New Mexico achieved more than 1 MW, or megawatt, of continuous output power. Because the power is so high, it is operated only for seconds at a time, producing several MJ of energy in an outburst.

Figure 5.2 is a photo of a laser in laboratory use, Figure 5.3 shows the standard warning sign for laser operation, and Figure 5.4 presents a laser used for entertainment effects.

5.3   Categorization

Lasers are categorized based on their safety level, which is represented by a class number. Classification for continuous lasers can appear as follows:

- Class I is considered safe because the laser and light are usually enclosed (CD players).

- Class II is also considered safe because the blink reflex of the eye will prevent damage (lasers of up to 1 mW power).

Figure 5.2. Laser light emissions (Nd:YAG 532 nm) in research applications.

Figure 5.3. The standard sign warning of laser operation.

- Class IIIa/3R lasers are usually up to 5 mW and involve a small risk of eye damage as staring into such a beam for several seconds is likely to cause minor eye damage.

- Class IIIb/3B can cause immediate severe eye damage upon exposure (lasers of up to 500 mW, such as those in CD and DVD writers).

- Class IV/4 lasers are the most dangerous and can cause skin burns. Caution must be exercised even when dealing with scattered light. Most industrial and scientific lasers are in this class.

Figure 5.4. A lighting effect with lasers.

Another method for the categorization of lasers is based on the gain medium. Table 5.1 lists some of the best known lasers along with some important information about them.

Table 5.2 illustrates that the efficiency of lasers is usually too low with the exception of carbon dioxide lasers, which are used in the industry for material processing.

5.4 Excimer Lasers

For most lasers, light is generated by means already described, such as the flow of electricity in semiconductors and electric discharges in a gas. Another type of laser is called the *excimer laser*, which is based on chemi-luminescence. In excimer lasers, a chemical reaction takes place where the product of the reaction is at a higher energy state than the reactants (population inversion).

Noble gases such as krypton and xenon do not readily form chemical compounds but when their atoms are excited (as during an electrical discharge) pseudo-molecules are formed of two noble gas atoms (dimers) or a noble gas atom and a halogen atom (complexes) such as fluorine or chlorine. The pseudo-molecules under excitation (excited dimer = excimer or excited complex = exciplex) easily release the excess energy returning to an unstable low-energy level leading to their dissociation and release of free atoms again. These are the right conditions for population inversion, which is essential for the functioning of lasers. Depending on the type of pseudo-molecules, the laser is called an *excimer* or an *exciplex* laser. In the case of Xe2, the radiation emitted is at 172 and 175 nm (same process and wavelengths we find in low-pressure xenon lamps) while other known wavelengths of laser emissions are 351 nm (XeF), 308 nm (XeCl), and 248 nm (KrF). The same principle applies (without the flow of electricity) in the way light sticks work. These sticks used in emergency situations or just for decoration contain various chemical compounds isolated from each other. By breaking the stick, the different compounds mix and the result of each chemical reaction is the emission of light. This way of creating light is very inefficient (about 1%) but

Table 5.1. Various Types of Lasers with Emissions in the Visible Part of the Spectrum and Their Applications

Gain Medium and Type of Laser	Wavelength/nm	Pumping Method	Applications
Helium-Neon/Gas	543, 632.8	Electrical discharge	Holography, spectroscopy, projections, shows, and displays
Argon ion/Gas	450–530 (488 and 514.5 most intense)	Electrical discharge	Lithography, projections, shows, and displays
Krypton ion/Gas	335–800	Electrical discharge	Scientific research, white lasers, projections, shows, and displays
Xenon ion/Gas	Multiple emission lines in UV-Vis-IR	Electrical discharge	Scientific research
Dyelasers	300–1000	Laser, discharge lamp	Spectroscopy, skin treatment
HeCd Laser metallic vapors	440, 325	Electrical discharge	Scientific research
Ruby/solid state	694.3	Discharge lamp	Holography, tattoo removal
Nd:YAG/solid state	1064, 532	Discharge lamp, diode laser	Material processing, medical procedures, scientific research, pumping other lasers, projections, shows, and displays
Ti:sapphire/solid state	650–1100	Laser	Spectroscopy, LIDAR (LIght Detection And Ranging)
Cr:Chrysoberyl (Alexandrite)/solid state	700–820	Discharge lamp, diode laser	Skin treatment, LIDAR
Semiconductor diode laser	Multiple emission lines in Vis-IR	Electric current	Telecommunications, printing, holography, 780 nm AlGaAs for CD scanning, projections, shows, and displays

Table 5.2. The Efficiency of the Laser Is Usually Too Low with the Exception of Carbon Dioxide Lasers That Are Used in the Industry for Material Processing

Laser	Efficiency/%
Argon ion	0.001–0.01
Carbon dioxide	5.0–20.0
Excimer	1.5–2.0
GaAlAs semiconductor	1.0–10
Helium-Neon	0.01–0.1
Nd: YAG	0.1–1.0
Ruby	0.1–1.0

in various enzymatic reactions taking place in living organisms such as fireflies, the efficiency can reach 90%.

5.5   Applications

The technology of lasers is applied in so many sections of human activity that it is impossible to imagine life without them. Laser light through optical fibers carry tremendous amounts of information (such as telephone conversations and computer connections). The telecommunication speeds of today are largely due to laser technology. Supermarket checkout scanners, CDs, DVDs, laser holograms, and laser printers are just a few of the countless everyday technologies that rely on lasers. Industrial lasers cut, drill, and weld materials ranging from paper and cloth to diamonds and exotic alloys far more efficiently and precisely than metal tools. Used in millions of medical procedures every year, lasers reduce the need for general anaesthesia. The heat of the beam cauterizes tissue as it cuts, resulting in almost bloodless surgery and less infection. For example, detached retinas cause blindness in thousands of people each year. If caught early, a laser can "weld" the retina back in place before permanent damage results. Before any other application, lasers were used for scientific research. At first, like masers, they were used to study atomic physics and chemistry. But uses were soon found in many fields. For example, focused laser beams are used as *optical tweezers* to manipulate biological samples such as red blood cells and microorganisms. Lasers can cool and trap atoms to create a strange new state of matter (the Bose Einstein condensate) that probes the most fundamental physics. Over the long run, none of the uses of lasers is likely to be more important than their help in making new discoveries, with unforeseeable uses of their own.

Bibliography

Silfvast, William T. (1996). *Laser Fundamentals.* Cambridge: Cambridge University Press.

Townes, Charles H. (1999). *How the Laser Happened: Adventures of a Scientist.* Oxford: Oxford University Press.

Wilson, J. and Hawkes, J.F.B. (1987). *Lasers: Principles and Applications.* Upper Saddle River, NJ: Prentice Hall.

6 The Technologies

Figure 6.1 shows all existing light sources and commercial lamps, which fall into three main technology families as introduced in the previous chapters: incandescent, discharge (plasma), and solid-state sources. Before opening the discussion on application, it is important to filter the various types of light sources as those with the greatest availability and circulation in the market can provide realistic and easy options for the readers. For this reason, two categorizations have been developed: primary (proposed for various applications) and secondary technologies (not included in the proposals). This chapter will begin with the former.

6.1 Technologies

The list of lamp technologies discussed and proposed for the situations dealt with in this book include:

- Incandescent halogen lamps

- Low- and high-pressure sodium lamps

- Fluorescent (including compact—CFL and inductive)

- Metal halide lamps (including ceramic versions)

- Light-emitting diodes

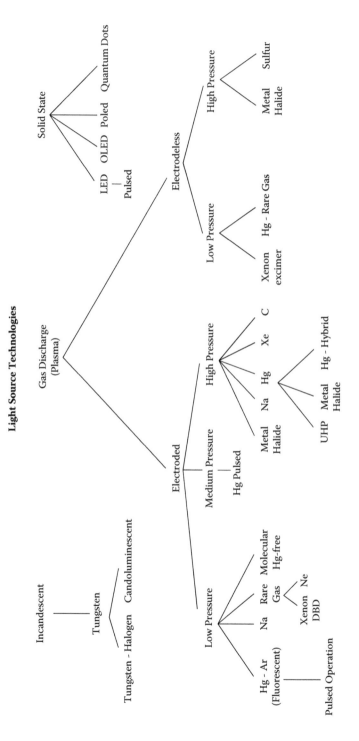

Figure 6.1. Categorization of lamp families.

Light sources are described using the following list of parameters:

- Efficacy

- Range of power—voltage

- Range of color temperatures

- Color rendering Index

- Lifetime

- Instant start

- Switching frequency wear

- Cost

- Dimming

- Mode of operation

- Extra gear

- Toxic/harmful materials

- Variety of shapes/geometries

The focus here is only on those parameters that are related to the light emitted (flux as seen from efficacy and power values, color temperature, and color rendering index) and service lifetime of the lamp. Occasionally, if deemed necessary, more details may be provided for specific cases.

It goes without saying that the technologies suggested due to their light quality must then be prioritized by the user based on other parameters and needs such as cost or mode of operation, but this is not dealt with here.

Table 6.1 through Table 6.3 present the parameter values for the lamp technologies that will be discussed. For convenience, the following acronyms are used:

- Halogen = HA

- Fluorescent = FL

- Low-pressure sodium = LPS

- High-pressure sodium = HPS

- Metal halide = MH

Table 6.1. Comparison of Key Characteristics and Parameter Values for Commercial Lamps

	HA	FL	LPS	HPS	MH	CMH	LED
Efficacy lm/W	30	<120	200	50–150	100	<95	<80
Power/W	<2000	5–165	<180	35–1000	<2000	20–250	0.1–7
Color temperature/K	<3500	Wide	1700	<3500	Wide	3000–4200	Wide
Color rendering index (CRI)	100	>90	0	20–85	>90	>90	>90
Lifetime/kHrs	2–5	10–30*	20	10–30	10–20	10–20	>50

* Inductive mercury fluorescent lamps offer more than a 60,000 hour lifetime.

Table 6.2. Comparison of Key Characteristics and Parameter Values for Commercial Lamps

	HA	FL	MH	LED
Efficacy	Low	High	High	Medium
CT	Warm	Wide	Wide	Wide
CRI	Highest	High	High	High
Lifetime	Short	Medium	Medium	Long

Table 6.3. Comparison of Key Characteristics and Parameter Values for Commercial Lamps

	LPS	HPS	CMH
Efficacy	High	High	High
CT	Warm	Warm	Medium
CRI	Lowest	Medium	High
Lifetime	Medium	Medium	Medium

Note: CT = Color temperature; CRI = Color rendering index.

- Ceramic metal halide = CMH

- Light-emitting diode = LED

6.1.1   Halogen Lamps

Halogen lamps (also see Chapter 2) offer all the features of incandescent lamps, which include:

- Excellent color rendering properties due to the continuous emission spectrum

- A warm white light

- IR emissions for some applications

- Instant starting

- High fluxes for high powers

6.1.2   Low- and High-Pressure Sodium Lamps

The great advantage of this family of lamps (also see Chapter 3) is the great efficacies they offer but other than that these lamps are rarely used due to their low color
rendering indices. As a result of their low color rendering indices combined with
their record high efficacies, these lamps were dominant for decades in applications
where high flux and cost efficiency were important but not color discrimination,
such as road/street lighting and other outdoor spaces such as parking lots, squares,
warehouses, and so on.

- Record high efficacies

- Poor color rendering properties

- High fluxes

- Warm white emissions from the high-pressure version

6.1.3   Fluorescent (Including Compact) (CFL)

Fluorescent lamps, which also represent compact lamps (also see Chapter 3) and
inductive mercury lamps, offer the following:

- A very wide range of color temperatures and colors

- Very good color rendering properties if the proper phosphor type is chosen

- Longer lifetimes than halogen lamps

- Very high efficacies

Fluorescent lamps suffer from switching electrode wear so an extra mechanism warms the electrodes in order to protect them but more energy is consumed this way. For this reason, cold cathode fluorescent lamps came onto the
market—they save energy but are supposed to be used in situations where the
lamp is not switched frequently and where the lamp is operated for long periods of
time. The other option when a fluorescent lamp is desired but duration is an issue
is to use the inductive versions that last up to four times longer.

The mercury inductive fluorescent (low-pressure discharge) source offers
fewer features than ordinary fluorescent lamps (less options for power, color temperature, and color rendering) and is a more expensive system. The big advantage
of this electrodeless technology is the significantly longer lifetime (the lifetime of
the lamp is essentially the lifetime of the gear powering it), so it may be an option
in situations where lamp replacement is difficult or costly. However, as a regular
option, we will discuss ordinary fluorescent lamps unless stated otherwise.

6.1.4   Metal Halide Lamps

Metal halide lamps (also see Chapter 3), which also include the ceramic metal halide lamps, offer the following:

- A wide range of color temperatures but mainly neutral and cool white tones

- High fluxes

- High efficacies

- Longer than halogen lifetimes

6.1.5   Light-Emitting Diodes

Light-emitting diode (LED) technology offers (also see Chapter 4) the following:

- Very long lifetimes

- Wide range of colors and color temperatures (ideal for dynamic lighting) but mainly neutral and cool white tones

- Good color rendering properties

- Low-power consumption

- Instant starting and frequent switching

These five technologies cover the vast majority of applications where a wide range of color temperatures and good color rendering is required. If in some applications one of these technologies is not ideal then a type from the excluded lamps will be discussed.

Table 6.4 through Table 6.6 present the recommended values of illuminance for a variety of tasks and environments, and Table 6.7 shows the color rendering

Table 6.4. Illuminance Values for a Range of Spaces and Tasks

Lux (Illuminance)	Type of Environment or Activity
5–20	Residential side street
10–50	Main road lighting
20–50	Outdoor workspace
30–150	Short stay rooms
100–200	Workrooms with noncontinuous usage
200–500	Work with simple visual requirements
300–750	Work with average visual requirements
500–1000	Work with high visual requirements
750–1500	Work with very high visual requirements
1000–2000	Work with special visual requirements
>2000	Extremely visually accurate work

Table 6.5. Illuminance Values for a Range of Spaces and Tasks

Lux (Illuminance)	Type of Outdoor Space
1–10	General storage spaces, security
10–50	Car parking spaces, cargo transfers
50–500	Sales, sports events, advertisement
500–1000	Sports events with audiences
1000–2000	Sports events with television coverage

Table 6.6. Illuminance Values for a Range of Spaces and Tasks

Conservation Category	Maximum Surface Illuminance (Lux)	Maximum Annual Exposure (Lux-Hours)
Objects without photosensitivity (metals, jewels, glass)	No limit	No limit
Average photosensitivity (wood, paintings, bones, skin)	200	600,000
High photosensitivity (fabrics, prints, drawings, watercolors, botanical samples)	50	150,000

Table 6.7. Color Rendering Index Requirements for a Range of Situations

Color Rendering Index (R_a)	Type of Environment or Activity
>90	Galleries, printing facilities
80–90	Homes, restaurants, fabric industry, museums
60–80	Offices, schools, light industry
40–60	Heavy industry, corridors, stairs
20–40	Outdoors

index requirements for certain situations. A more detailed discussion can be found in Chapter 8 ("Illuminating Spaces") on indoor and outdoor lighting proposals.

6.2   Secondary Technologies

There are a large number of lamps that have not been able to make an impact in any lighting application; they are not considered readily available or are being phased out for various reasons. These lamps will be considered as secondary technologies and other lamp types will be proposed if they fit the need. Examples of lamps that generally will not be discussed are as follows, with a brief overview of each following the list:

- Organic light-emitting diodes (OLEDs)
- Conventional incandescent lamps
- Sulfur lamps
- Mercury vapor discharge lamps

- Super-high-pressure mercury discharge lamps
- Xenon excimer and high-pressure lamps
- Carbon arc lamps
- Neon lamps
- Ceramic metal halide lamps
- Inductive mercury fluorescent lamps

6.2.1   Organic Light-Emitting Diodes

This technology is excluded as it is still under development and has not really penetrated the lighting market (thus far these are used as small displays).

6.2.2   Conventional Incandescent Lamps

Most conventional incandescent lamps are gradually being phased out due to energy savings concerns but the technology of incandescence will be represented by halogen incandescent lamps, which share the same properties.

6.2.3   Sulfur Lamps

These are lamps that have an inductive sulfur discharge. Despite its attractive emission spectrum that resembles the eye sensitivity curve, this lamp has not been a commercial success so it will not be included due to lack of availability.

6.2.4   Mercury Vapor Discharge Lamps

These are lamps that have medium to high pressure including hybrids, with mercury plus incandescence coming from the resistor ballast. The mercury lamp technology is an old technology from which the metal halide lamps developed as more metal salts were being added to mercury in order to create a range of spectra and wavelength emissions. With the phasing out of mercury, this technology has essentially been replaced by the metal halide lamps.

6.2.5   Super-High-Pressure Mercury Discharge Lamps

This is a special type of lamp that is used in specific applications such as projection systems.

6.2.6   Xenon Excimer and High-Pressure Lamps

Xenon-based lamps are low efficacy lamps used for specific applications such as flash photography, photocopy machines, and studio lighting but not for general lighting where other technologies can be used. This lack of wide availability, the more complicated gear, and not so significant penetration does not qualify them as first choices in general but will be mentioned in some cases.

6.2.7   Carbon Arc Lamps

Carbon arc lamps are an obsolete technology, which are not used for any general lighting applications.

6.2.8 Neon Lamps

This discharge lamp has for decades had a major and dominant role in signage and decoration. However, this discharge technology is being displaced by LEDs in those markets.

6.2.9 Ceramic Metal Halide Lamps

These lamps will be represented by the metal halide lamps, in general.

6.2.10 Inductive Mercury Fluorescent Lamps

These will be represented by fluorescent lamps. Although the inductive versions offer less flexibility on several features, they have the great advantage of much longer lifetimes and lack of switching wear so in some cases they will be discussed and proposed. Table 6.8 through Table 6.11 show the characteristics for some of the above-mentioned technologies.

Table 6.8. Some of the Characteristics of Lamp Technologies That Will Be Excluded from Discussions or Mentioned in Some Special Applications Only

	Incandescent	CFL
Efficacy lm/W	20	65
Power/W	15–1000	5–55
Color temperature/K	2800	Wide
CRI	100	85
Lifetime/hours	1000	1000

Table 6.9. Some of the Characteristics of Lamp Technologies That Will Be Excluded from Discussions or Mentioned in Some Special Applications Only

	Mercury Hybrid	Sulfur
Efficacy lm/W	20	95
Power/W	100–500	1000–6000
Color temperature/K	3300–3700	6000
CRI	50–70	80
Lifetime/hours	10,000	20,000

Table 6.10. Some of the Characteristics of Lamp Technologies That Will Be Excluded from Discussions or Mentioned in Some Special Applications Only

	Hg High Pressure	Hg Very High Pressure
Efficacy lm/W	60	60
Power/W	50–1000	100–250
Color temperature/K	3000–4000	7500
CRI	15–55	<60
Lifetime/hours	10,000–30,000	10,000

Table 6.11. Some of the Characteristics of Lamp Technologies That Will Be Excluded from Discussions or Mentioned in Some Special Applications Only

	Xe Excimer	Xe High Pressure
Efficacy lm/W	30	>30
Power/W	20–130	1–15 kW
Color temperature/K	8000	>6000
CRI	85	>90
Lifetime/hours	100,000	2000

7 Luminaires

The lamp is the heart of every lighting fixture but a complete luminaire usually includes most of the following: reflectors for directing the light, optics for diffusing the light, an aperture (with or without a lens), the housing for lamp alignment and protection, ballast, if required, and connection to a power source. There is a great variety of luminaires and choices are made depending on the lamp to be used, but also with aesthetic criteria. Figures 7.1 and 7.2 show two typical luminaires found in most house and office ceilings, respectively. For incandescent lamps, with the exception of spot halogen lamps, all luminaires have a decorative design and character. Various types of reflectors are used in the design of luminaires, such as flat, elliptical, parabolic, spherical, and a combination of these.

7.1 Key Considerations

For fluorescent lamps, the luminaires not only employ reflectors so that light is directed toward the desired work or living space but they are also designed in such a way that the generated heat is dissipated and carried away. For both high- and low-pressure discharge lamps, the luminaires must also be able to accommodate the electronic gear.

When choosing a luminaire one has to take into account its efficiency, which is defined as the ratio of the luminous flux of the luminaire over the luminous flux of the lamp(s). This is also known as the *light output ratio* (LOR). Another coefficient is utilization (CU), which is a measure of the efficiency of a luminaire's transfer of luminous energy to the working plane in a particular area. A CU measures the light that actually reaches the desired plane as a percentage of the total light produced by the fixture.

Figure 7.1. A general purpose luminaire.

Figure 7.2. A luminaire for linear fluorescent tubes.

The lamps that dominate the entertainment industry and displays are those with high power and white light emission, such as incandescent/halogen lamps, xenon high-pressure lamps, and in some cases, carbon arc lamps. For the emission of colored light, various filters are used in front of the light source, although with the ever-increasing use of LEDs, filters are no longer needed.

7.2 Categories

Luminaires can be broadly separated into two categories: *floodlights*, which illuminate a wide area (diffuse light) such as the PAR luminaires (parabolic

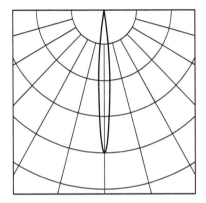

Figure 7.3. Characteristic beam profiles for spotlights.

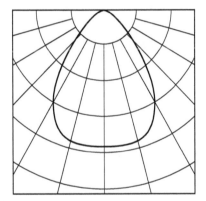

Figure 7.4. Characteristic beam profiles for floodlights.

aluminized reflector), and *spotlights* (sometimes known as *profiles*), which produce a narrower, more controllable light beam (focused light). The diagrams in Figures 7.3 and 7.4 are examples of the radial distribution of the emitted light in each case.

Floodlight luminaires are also categorized as *circular*, which suit incandescent lamps best (their diameter is proportional to the wattage of the light source used), and *rectangular*, which suit high-pressure discharge lamps best (see Figures 7.5 and 7.6).

Luminaires are also categorized depending on whether the luminaire is totally exposed or concealed behind a ceiling or wall. The ceiling-mounted version is often called a *downlight*.

Some examples of recessed fixtures include *cans*, which is a general term for inexpensive downlighting products recessed into the ceiling, and *troffer luminaires*, which refer to recessed fluorescent lights (the word comes from the combination of *trough* and *coffer*). Examples of exposed luminaires include the chandelier, pendant lights which are suspended from a ceiling usually with a chain (Figure 7.7); undercabinet lights which are mounted below kitchen wall

Figure 7.5. A floodlight luminaire for high-pressure discharge lamps.

Figure 7.6. Another floodlight luminaire for high-pressure discharge lamps.

cabinets; high and low bay lights typically used for general lighting in industrial buildings; and strip lights which are usually long lines of fluorescent lamps used in a warehouse or factory.

There is of course a whole range of luminaires used in outdoor lighting to illuminate streets, parking lots, building exteriors (Figure 7.8), and architectural

Figure 7.7. Exposed pendant lights.

Figure 7.8. Luminaire used in outdoor lighting and specifically for building exteriors.

details. These luminaires can be pole mounted to illuminate streets and parking lots or mounted in the ground at low levels for illuminating walkways.

Some of the most expensive and sophisticated lighting systems are those used in professional applications such as stage lighting. Here, again, the categorization takes place depending on whether the luminaire is used for floodlighting or as a spotlight. Some widely used floodlight systems are discussed next.

The parabolic aluminized reflector lights, or PAR lights, or PAR cans, are nonfocusable instruments. PAR cans consist mainly of a metal cylinder with a sealed-beam parabolic reflector lamp at one end. Like an old-fashioned automotive headlight, the reflector is integral to the lamp and the beam spread of the unit is not adjustable except by changing the lamp. The instrument throws an unfocused beam, the shape of which depends on what type of lamp is in the instrument. Color frames can be utilized with most cans by using the clips that are present on the front of the instrument. PAR lights are low in cost, lightweight, easy to maintain, and are highly durable. See Figure 7.9 for an example.

Strip lights, also known as *cyclorama,* are long housings that typically contain multiple lamps arranged along the length of the instrument and emit light perpendicular to its length. Lamps are often covered with individual filters of multiple colors (usually the primary colors) that are controlled by a separate electrical dimmer circuit. Varying the intensity of the different colors allows for dynamic lighting.

The ellipsoidal reflector floodlights, better known as *scoop* lights, are circular fixtures that do not have any lenses. They have an incandescent lamp mounted inside a large parabolic metal reflector that directs the light out of the fixture. Since they do not have any sort of lens system they are cheaper than other fixtures. However, the downside is that the light cannot be focused at all (even PARs allow

Figure 7.9. Parabolic aluminized reflector.

more control than scoops). Scoops can accommodate filters for tailoring the emitted spectrum.

Some known spotlight categories that are used widely in the entertainment industry are Fresnel spotlights and the ellipsoidal reflector spotlights.

Fresnel spotlights consist of a mirrored parabolic reflector, an incandescent lamp at the focus, and a round Fresnel lens. The distinctive lens has a *stepped* appearance instead of the *full* or *smooth* appearance like those used in other lanterns, and it was actually built this way so that lighthouses could throw light farther. It is this lens that lends the instrument both its name and its soft-focus characteristic. Adjustments allow for the focusing of the beam from spot to flood. Provisions for color frames are generally present on the front of Fresnels. The lamp and reflector remain as a fixed unit inside the housing, and are moved back and forth to focus the light.

The ellipsoidal reflector spotlight, ERS, or profile (UK) is a common nonautomated luminaire which usually consists of an incandescent lamp, an elliptical reflector, and convex lenses. The ERS (Figure 7.10) is a flexible instrument that allows for the focus and shaping of the beam. Various colors and patterns can also be produced with appropriate filters and shutters. Large versions are used as followspots to pick out specific people or objects on a stage.

Finally, the use of automatic *moving lights* (DMX is the protocol for the electronic control of lighting), as seen in Figure 7.11, is becoming increasingly

Figure 7.10. Ellipsoidal reflector spotlight (ERS).

Figure 7.11. Smart lighting with total electronic control.

widespread. The direction of the light beam is controlled by moving the luminaire, or just the reflectors and lenses, which can rotate and tilt. All functions can be computer and remote controlled. Such instruments provide features such as color changing, pattern changing, tilting, strobing, and so forth. These are the most complicated instruments available, and involve a considerable number of technologies. There are many designs of automated lighting systems from many companies.

8 Illuminating Spaces

In this chapter, we will look at some representative cases of indoor and outdoor lighting. It is impossible to cover every single space or room that someone would want to illuminate, therefore the most common ones have been selected in the hopes that the principles by which light source selections are made will act as a guide to the reader for any situation. The focus is mainly on color temperatures and color rendering options while the levels of illumination should be decided by the designer based on existing standards and various parameters such as availability of natural light, the age of the people occupying the space, the nature of the tasks performed, the room area and ceiling height, and so on.

8.1 Indoor Lighting

8.1.1 Home Lighting

Lighting at home should be characterized by the relaxed and cozy feeling that warm white sources give and also by good color rendering (it is important to be able to see the color of your chosen decoration, your clothes, the food you prepare, etc.).

A lamp that can satisfy both these requirements is the *halogen lamp* and it is for this reason that homes around the world are dominated by incandescent lighting. The cost of the bulbs is another reason, but as more and more people focus on energy savings, the options of fluorescent sources and light-emitting diodes (LEDs) are becoming more popular.

The advantage of LEDs is the long lifetime, the flexibility in geometries and shapes for decorative lighting, and the ability to create dynamic lighting and a range of atmospheres and colors.

Let's take a look at some different rooms.

Kitchen: Food requires a high color rendering source so halogen lamps or LED and fluorescent lamps with high CRI are recommended (fluorescent lamps with a silicon coating for containment of toxic materials in case of breakage are available).

Living room: This is an area where people want a relaxed environment so a warm source is needed. Halogen lamps are usually chosen but warm white fluorescent lamps are also suitable. For more atmosphere and dynamic lighting, LED systems are suggested.

Bedroom: This is a room where a warm source is needed to create a relaxed environment so halogen lamps are usually chosen but warm white fluorescent and warm white LED lamps can also be used.

Study room: This is a room where one should be able to focus so cool white fluorescent lamps or cool white LEDs are proposed.

Storage room: Color rendering or high luminance or specific color temperature is not important here so unless other issues point to a specific lamp then LED lighting which offers a long lifetime is the proposed choice.

Corridors/stairs: Color rendering or high luminance or specific color temperature is also not important here but instant start, frequent switching, and long lifetime are, so unless other issues point to a specific lamp then again LED lighting would be the proposed choice.

8.1.2 Bathrooms

When it comes to bathrooms there are two opposite options. One can choose halogen lamps that create a relaxed atmosphere and also heat up the space or one can choose cool white sources (cool fluorescent or LED lamps) that give the impression of cleanliness.

8.1.3 Schools

Classrooms are places where one wants to feel stimulated but also relaxed. Therefore, one should compromise between warm and cool white sources. LED systems these days offer dynamic lighting and simulation of daylight changes but assuming that these would be very expensive to fit in schools fluorescent or metal halide lamps with a neutral white color temperature of around 4500 K are proposed. If one uses fluorescent lamps then the **inductive** version would be a good choice due to the longer lifetime they offer.

8.1.4 Restaurants

Restaurants have two requirements: they need a good color rendering (food should always be color rendered accurately) and a cozy and relaxed atmosphere. This is why halogen lamps are suitable and more widely used. An additional advantage is the lack of toxic materials in these lamps. As a second option, LED systems that offer dynamic and varying atmospheres as well as decorative lighting should be considered.

8.1.5 Offices

Offices are also discussed in Section 8.2 on dynamic lighting. Depending on the task or time of the day, workers may need warm or cool white light. Certain systems with fluorescent lamps or LEDs offer this dynamic solution.

There is a special case of a fluorescent lamp found on the Japanese market (Figure 8.1) that offers the same features as any other having the additional feature of phosphorescence. This means that the phosphor coated on the inside walls not only converts the UV emissions of mercury into visible light but it also phosphoresces for up to 30 minutes after it is switched off. This particular lamp is not widely available but if the technology to be used is fluorescent lamps and if one could have access to this particular product, then it would be of great value when it comes to the safety of certain buildings (such as schools, senior citizen centers, offices, industry, etc.). Without any additional safety circuits and systems installed, buildings equipped with such lamps would give the occupants the required light to evacuate in case of emergency.

Figure 8.1. A special fluorescent lamp that also phosphoresces after it is switched off.

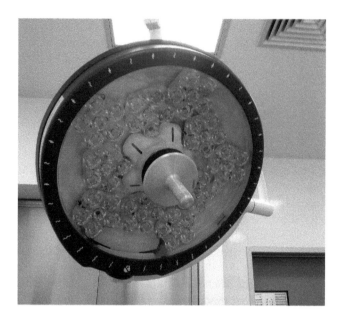

Figure 8.2. LED-based products have started being used in operating rooms.

8.1.6 Hospitals

When one discusses hospitals, a distinction should be made between the diagnostic rooms (Figure 8.2) and the rooms where visitors wait and patients rest or heal. In the first case, the rendering index is important so halogen lamps can be used but so can fluorescent and LED lamps as long as their CRI is very high. A doctor after all must be able to tell if the color of the patient's skin looks healthy. In the other cases, the priority is on the color temperature that should create a relaxing environment. Warm white light sources such as halogen should be used or fluorescent and LED lamps of similar color temperatures (<3500 K) should be used. The healing aspect of patients and light therapy will be discussed in more detail in Chapter 9, "Light Sources and Health."

8.1.7 Dentistry

A dentist working on people's teeth must certainly have an illuminated working space with sources of the best color rendering properties in order to be able to make the best color matching identifications. Halogen lamps and special high CRI LEDs or fluorescents are recommended. The waiting rooms should, of course, be illuminated with warm white sources to enhance relaxation, so halogen lamps or warm white fluorescent and LED lamps are again proposed.

8.1.8 Churches and Places of Worship

Places of worship are obviously spaces where people look for a calm environment. The warm white light of halogens and warm white fluorescent or LED lamps will provide the appropriate atmosphere.

8.1.9   Senior Citizen Homes

This is clearly a place where relaxation and coziness is desired so warm white sources such as halogen lamps or warm white fluorescent and LED lamps should be chosen. Of course, when it comes to senior citizens there is a range of needs and health conditions to deal with (also see Chapter 9, "Light Sources and Health").

8.1.10   Gym and Exercise Rooms

Cool white light promotes focus and stimulation so fluorescent and LED lamps with cool white light emissions should be used in this case.

8.1.11   Hotels

Hotels and other accommodation spaces have requirements that do not differ much from home lighting. As the objective is again to have relaxing and cozy atmospheres, the halogen and warm white fluorescent or LED lamps should be used. LED systems may be used for decorative lighting and the creation of fun and dynamic illumination.

8.1.12   Theaters and Concert Halls

All entertainment spaces should be illuminated with warm white sources in order to create a cozy and relaxed atmosphere. Warm white fluorescent and LED lamps as well as halogen lamps are recommended.

8.1.13   Industry

Depending on the type of industry one might want to use metal halide lamps for high illumination levels and decent color rendering, fluorescent lamps for better color rendering, or inductive fluorescent lamps for good rendering and a longer lifetime. LED systems with good color rendering are also an option if cost is not prohibitive and if the illumination levels are sufficient.

8.1.14   Warehouses

A warehouse has no special requirements for color rendering or color temperature and any light source technology will work, so it comes down to cost and availability. Many warehouses use sodium lamps that provide high efficacies but low color rendering or metal halide lamps with good efficacies and better color rendering.

8.1.15   Ships/Airplanes

Most areas have already been covered (kitchens, sleep areas, long trips requiring dynamic lighting, etc.). For areas that have no specific requirements, but only durability and good performance in cold or harsh weather conditions, then LEDs are the proposed solution.

8.1.16   Shops

This is a general category of spaces and the answer to the question of what kind of lamp and color temperature to use is "it depends on what the shop sells." Lighting is meant to enhance the image of the product and create the appropriate atmosphere for this. Generally speaking, halogen lamps are used as spotlights to highlight the products and create warm atmospheres if desired. Fluorescent and LED lamps can be used for both warm and cool white light illuminations. A shop selling coats would want to use a cool white source to create the feeling of winter and a shop selling gold would want to use warm tones to enhance its product.

Generally, a light source with good color rendering is desired especially if one of the important features of the product is its color (such as food or flowers). A butcher's shop would need a fluorescent lamp with emphasis in red emissions in order to enhance the color of the blood, hence the freshness, of the meat it sells. However, when dealing with low temperature conditions such as fridges or freezers, then cool white sources should be preferred. Cool white fluorescent lamps can be used as well as cool white LEDs (LED technology is seriously limited by thermal management issues so a low temperature environment would be ideal for it).

8.1.17 Emergency Lights

When it comes to emergency lighting the most important feature is reliability, which means it should have a long lifetime, be instant starting, flexible, and be resilient to damage. Although many sources could be used, LEDs seem to satisfy most criteria (see Figures 8.3 and 8.4). It is important to ensure that the whole system does not fail if one LED unit fails—therefore circuitry and connections are critical here.

As for the color of emergency lights, it depends on where they will be used. The human eye's photopic sensitivity curve peaks at 555 nm so a green or yellow light will be more easily detected with foveal vision in bright conditions. In dark conditions, eye sensitivity shifts to the blue region (peak at 507 nm), so blue lights might be more detectable and also favored by peripheral vision (rods). If smoke or particles that might scatter light are involved, then longer wavelengths (yellow or red) are the preference as light scatters less in these conditions. Once again it has

Figure 8.3. Emergency lights of specific colors (such as red). LEDs have started to dominate this application too.

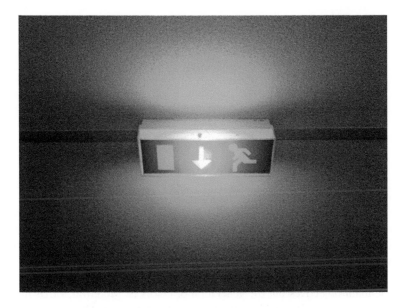

Figure 8.4. Emergency lights of specific colors (such as green).

been shown that based on a small number of principles one can make the decision on the appropriate sources and light wavelengths for a variety of situations.

8.1.18   Lamps for Horticulture and Aquariums

For such applications as horticulture and aquariums where the green and blue wavelengths are of importance, then neutral and cool white sources of relatively high CRI should be used. The first choices should be cool white fluorescent and metal halide lamps. Once again, LED systems with the appropriate color temperatures and rendering indexes can be used if cost is not an obstacle.

8.1.19   High Color Rendering Applications

When color rendering is of the utmost importance then the user can use the halogen lamp (that can also be used as a spotlight to highlight an object), which offers a color rendering index of 100, or fluorescent lamp that employs an appropriate phosphor to give it a color rendering as close to the optimum as possible (>95) or high CRI LEDs. However, infrared emitting halogen lamps and UV emitting halogen or fluorescent lamps should be avoided or used with filters if the illuminated object is sensitive to this kind of radiation. Even the visible light levels should be carefully controlled in order to avoid deterioration of certain materials.

There are certain applications and work environments where color rendering is absolutely necessary. Some of these are

- Galleries

- Fabric industry

- Art studios

- Printing facilities

- Drawing offices

- Florists

- Make-up rooms

- Museums

8.1.20 Signs and Decoration

The industry of luminous signs and decorative lights has traditionally been an area dominated by neon lamps and other low-pressure discharge lamps (the mercury discharge lamps can generate any color with the use of the appropriate phosphors). Now, however, the technology of LEDs has taken over. In fact, this was the first area where LEDs penetrated the market, as lumens and efficacy are not the main objectives but the generation of colors is. Moreover, LEDs have a low-power consumption and long lifetime (see Figures 8.5 and 8.6).

8.1.21 Studio and Stage Lighting

Lighting is one of the most important elements in filmmaking, theater, and studio photography. Lighting can affect the entire mood of a scene as well as enhance specific feelings or messages. The most usual terms in film and studio lighting are *hard* and *soft* lighting.

A hard light is usually a small intense light that creates sharp shadows. A hard light is more directional and can create a dramatic mood. A soft light is usually a larger and diffuse light that creates undefined shadows. A hard light can be converted into a soft light with the use of a panel to bounce the light. Also, a panel in front of the source can make the shadow more diffuse.

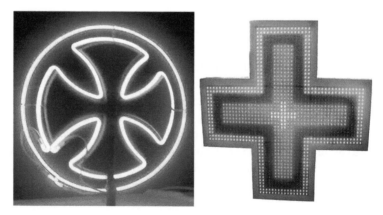

Figure 8.5. Neon and mercury discharges (left) are being displaced by LED systems in the signage industry.

Figure 8.6. LED strips of any color are commonly used for decorative lighting.

Usually, in studios, technicians use the three-point lighting technique. The three lights needed for a scene are:

1. A *key light* (hard light) at the front and slightly to the right or left to illuminate the central subject.

2. A *fill light* (a diffuse light to the left or right so as to balance the shadows made by the key light).

3. A *back light* (hard light) to give the impression of a three-dimensional entity.

Sometimes a fourth light is used to illuminate the background.

With respect to the intensity of the light sources, less light gives a smaller depth of field. One can control the intensity by spotting or flooding the lamp, by moving the lamp or the subject, by attaching a dimmer, or by blocking it with a filter (such as a neutral density filter). The direction of the light can also suggest a season or a time of day.

Regarding the color of the light sources, there are two main types used: cool white sources at higher than 5000 K so that sunlight at midday is simulated and warm white sources at temperatures lower than 3500 K.

Tungsten halogen lamps are used in the majority of theatrical and studio (film and television) fixtures, including ellipsoidal reflector spotlights and parabolic aluminized reflector cans. These warm white light sources have excellent color rendering properties but one can also change the light emitted with the use of panels (hard or soft) and filters.

A *high-pressure sodium lamp* could also be used as a warm source offering high levels of illuminance, although the color rendering would not be exceptional.

If one wants to use cool white light sources to simulate daylight at midday or moonlight then the options, besides halogen lamps with filters, are metal halide lamps with high CRI and, traditionally, high-pressure xenon lamps have also been used in studios.

As LED systems are being developed to operate at higher powers and fluxes they are being used more often in stage lighting due to the flexibility they offer in scene settings by changing colors and white color temperatures. Figure 8.7 through Figure 8.9 present examples of how LED systems are used for a wide range of studio and stage applications (illuminating scenes, providing green light for computer-generated imaging, and as floodlights for larger areas).

An additional note should be mentioned for the field of photography. In addition to the sources discussed above that are used to illuminate a scene, photographers have to use two more illuminating systems: lamps for darkroom chemical film development and light sources for compact camera flash lights.

In both traditional and digital photography the flash lights of the cameras used to be low-pressure xenon sources but they are now being replaced by white LEDs (see Figure 8.10). In fact, all modern mobile phones employ LED light sources these days. In traditional photography, the darkrooms where film is developed are illuminated by red light sources (film is sensitive to shorter wavelengths), so red LED lamps can now be used (neon discharge lights emit in the red area of the visible spectrum but they are being replaced by LEDs in almost all applications).

More information on some of the luminaires used for stage lighting can be found in Chapter 7.

Figure 8.7. LED systems for scene lighting.

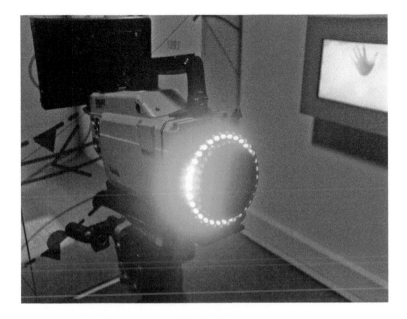

Figure 8.8. Computer-generated imaging (CGI) filming.

Figure 8.9. Stage lighting floodlight.

8.2   Dynamic Lighting

In Chapter 1 ("Basic Principles of Light and Vision") we discussed the human internal clock called the *circadian rhythm* and that it is regulated and affected by natural light changes which, in turn, vary the melatonin levels throughout the day. However, in modern societies, humans spend more time in artificial and indoor environments where natural light is absent and the lighting environment is simply static. It is important to try to recreate these natural rhythms of light in order to

Figure 8.10. An array of white LEDs for use as a photography flash lamp.

have healthy environments and in order to create the appropriate lighting environments to match several needs.

This attempt to manipulate the circadian rhythm of humans with the use of artificial light sources and to create any desired lighting environment at will is called *dynamic lighting*.

Dynamic lighting is:

- A creation of a proper atmosphere for a certain activity

- A creation of a proper atmosphere for a certain mood

- A simulation of natural light that is not static but changes during the day and generally over time

- A creation of different aesthetic effects

As discussed previously, a cool white light creates a stimulating and active environment (>5000°C), while a warm white light creates a relaxing and pleasurable environment (<3500°C). By looking at the available technologies we can make a list of cool white sources and warm ones.

High color temperature (cool) white light sources:

- Cool white fluorescent lamps (all types)

- Cool white LEDs

- Metal halide lamps

Low color temperature (warm) white light sources:

- Halogen lamps

- Warm white fluorescent lamps

- Warm white LEDs

There are some products on the market that provide the user with the option to control and change the color temperature (Figures 8.11 and 8.12). These products

Figure 8.11. An example of how dynamic lighting can be achieved with the use of multiple fluorescent lamps with different color temperatures.

Figure 8.12. Another example of dynamic lighting.

use multiple light sources with some emitting warm white light (i.e., 2700 K) and others a cool white light (i.e., 6500 K). By dimming different lamps the user can control the overall color temperature of the system. So, for example, a cool white light can be chosen during a task that requires focus, and a warm white light would create a more comfortable environment suitable for relaxation. With proper dimming the whole color temperature range can be achieved.

The most appropriate technologies for such a task (dynamic and continuously changing lighting environment) are fluorescent lamps and LEDs because not only have these lamps become the standard technology for indoor and office lighting, but also because these families of light sources come in different color temperature types by employing the appropriate phosphors. LEDs have an extra advantage due to their easy dimming features.

The disadvantage of using multiple light sources for the creation of varying lighting environments is that expensive optics are needed in order to mix and diffuse the light from each source and create a homogenous final emission. During the process of diffusion and mixing of light from fluorescent tubes in known products, up to 50% of the light is lost.

Moreover, by using three lamps that each emit light of one of the three primary colors (red, green, blue), then any other color can be created by mixing the primaries and space lighting can be accomplished in many ways. This is a usual practice for LED systems.

A popular application of systems for dynamic lighting is the simulation of daylight in spaces that have no access to natural light (no windows). Dynamic lighting systems can reproduce the changes in color temperature that take place during the day, creating a feeling of time passing and of natural changes. For natural/sun light transport to indoor spaces, the reader should also refer to light tube products which are not, however, discussed here.

Regardless of the number or types of light sources, dynamic lighting systems offer flexibility and numerous options. Whether preprogrammed or set by the user, dynamic lighting systems can accommodate many lighting needs and control the atmosphere at will. Nowadays, dynamic LED light sources can even be smartphone controlled (Figure 8.13).

The sections below provide some examples of different environments, functions, and atmospheres that dynamic lighting can create.

8.2.1   Work Environments

Meetings—Medium level of brightness and high color temperature for stimulation.

Focused work—High color temperature and high brightness levels for office work that demands focus.

Relaxation—Low brightness levels and low color temperature for resting or hosting guests.

Product promotion—Different tones of white light enhance the image of certain products so, for example, winter products appear more attractive under a cool white light, and so on.

Figure 8.13. Smartphone-controlled LED dynamic light sources.

8.2.2   Changes during the Workday

Morning—Higher color temperatures and brightness levels stimulate and boost activity.

Midday—Lowering brightness levels and color temperatures create a relaxing environment.

Afternoon—This is the part of day when most people feel tired. An increase in brightness levels and color temperatures lifts activity levels.

Evening—For those who are working late, a warm white light can create a pleasant atmosphere.

Daylight simulation—The mood of people working in closed spaces with no natural light (no windows) can be affected by monotonous static lighting. Dynamic lighting systems offer a solution by creating a changing atmosphere, simulating the changes of natural light during the day.

All these effects that cool and warm light sources can produce find uses beyond offices and homes and dynamic lighting. Or, simply the use of a specific source to enhance alertness or relaxation can be useful in a variety of environments and applications. Some examples where we can utilize more lamps (besides fluorescent and LEDs) that offer cool or warm white light are discussed next.

Figure 8.14. The use of LED systems in airplanes allows for dynamic lighting and more comfortable long haul flights.

8.2.3   Other Applications for Dynamic Lighting

Hospitals/nursing homes—Creating comfortable and relaxing environments that also assist in the healing process.

Night shifts—Creating stimulating environments to avoid fatigue and accidents.

Long haul flights—The appropriate tailoring of light variation to match the natural rhythm can deal with jet lag syndrome [1] (Figure 8.14).

Shops—Choosing the appropriate color temperature to accommodate a variety of products.

8.3   Outdoor Lighting

8.3.1   Open Spaces

When we discuss outdoor spaces, many different areas can be included, such as:

- Parking lots

- Streets and roads

- Stadiums

- Squares

- Bridges

In all these cases and more, the lighting systems that are used must meet a number of criteria such as:

- The illumination must be uniform.

- The luminaires and poles must be protected from weather conditions and vandalism.

- The light output/flux from the source must be high in order to illuminate the large spaces.

- As these systems operate throughout the night they must be cost effective so high efficacies and a long lifetime are required.

Figure 8.15 illustrates how an outdoor space such as a bridge may involve different technologies. In the photograph we see three different lighting schemes being employed: metal halides for the main road, sodium discharges on the sides for visibility from a distance, and projectors floodlighting the cables with blue light for decorative purposes.

Traditionally, for outdoor spaces a wide range of light source technologies have been used with sodium discharges (high and low pressure) being the most

Figure 8.15. An illuminated bridge using different lighting technologies.

frequent for efficacy reasons followed by mercury lamps. However, as the subject of outdoor lighting is closely connected to the safety and security of people (pedestrians and drivers), it is important to refer to the next section on lighting and security.

8.3.2 Lighting and Security

The issue of whether lighting can contribute to the reduction of crime has been the subject of many studies and is still under investigation [2–6].

There seems to be at least two sides to this argument and here are some of the points supported.

How improved lighting could reduce crime:

- Improved lighting deters potential offenders by increasing the risk that they will be seen or recognized when committing crimes.

- Increased lighting leads to more surveillance by the maintenance lighting workers.

- Police forces become more visible, thus leading to a decision to desist from crime.

- New lighting can encourage residents to walk late or spend more time in the area thus increasing informal surveillance.

- New lighting shows that the local authorities are determined to work on the problem of crime thus deterring possible offenders.

- Better lighting can increase community strength leading to a greater willingness to intervene in crime and to report it.

How improved lighting could increase crime:

- Increased social activity outside the home in the evenings can increase the number of unoccupied homes available for burglary.

- Increased visibility of potential victims allows better assessment of their vulnerability and the value of what they carry. Offenders might more easily be able to see if parked cars contain valuable items.

- Increased visibility allows better judgment of the proximity of "capable guardians" who might intervene in crime.

- Better lighting might facilitate activities such as drug dealing and prostitution.

- Better lit streets might attract disorderly youths from nearby areas.

- Improved lighting of rarely used footpaths might facilitate undesirable behavior.

- Glare can compromise the vision of civilians and criminals may exploit this.

- Glare can compromise the recording of closed circuit surveillance cameras.

The matter is still under research by some dedicated lighting experts that conduct experiments in various urban parts of the world or perform meta-analyses of existing data. When deciding that changes in lighting must be made, there are some important points that should be considered:

1. Lighting must be uniform to eliminate both glare and dark shadows (Figure 8.16). Uniformity is a measurement of how evenly light is being distributed across a designated area. Higher uniformity values mean that bright areas of illumination are not adjacent to patches of darkness as these are circumstances within which the human eye finds it very difficult to see clearly. In order to calculate the Uo value of a lighting system the minimum levels in lux are divided by the average levels in lux. A standard uniformity value is 40% or 0.4 Uo. If a lighting system has a minimum of 8 lux and an average of 20 lux, divide 8 by 20, which results in a value of 0.4, and that would be the recommended value.

Figure 8.16. A uniformly illuminated road where the traditional sodium lamps have been replaced by white light emitting systems.

2. Instead of a few powerful lamps, multiple lamps with moderate power will reduce glare, provide more even illumination with reduced pools of shadow, and provide some redundancy if a lamp fails.

3. The lighting systems must be designed in such a way (luminaire) to protect it from vandalism (height of pole, materials, and protection masks). Generally speaking, the higher up a lamp is fixed, the bigger the area of illumination but the greatest the requirement for lamps with greater lumen outputs.

4. The chosen luminaire must perform two important functions: Reduce glare and light pollution. Cutoff luminaires are becoming popular and are being promoted by many interested organized groups, such as astronomers and light pollution activists that also require outdoor lighting systems to be efficient and focused (Figures 8.17 and 8.18).

As for the type of lamps to be used, a number of proposals have been made for a number of different situations in an effort to match each technology's features to the appropriate conditions.

Off-axis visual detection is quite important in safety issues (criminals usually sneak up and do not approach head on), so light sources that stimulate the peripheral vision (rod stimulation) are important. This means that lamps with blue emissions should be preferred [7,8].

The same reasoning is applied to driving safety where obstacles and pedestrians can be off the central vision axis. Research shows that the illumination of roads and streets with sources that aid peripheral vision have proven to be beneficial. The fact that the eye functions in mesopic conditions further enhances this reasoning.

Moreover, a high color rendering index is desired as this will enhance color contrast (enhances perception of brightness) and gives accurate information on events (better recognition and surveillance).

Finally, as these lights usually remain on throughout the night, economic systems should be chosen (high flux, long lifetime, high efficiencies).

Based on the above points on color rendering, lifetime, more emissions in the blue region, and high fluxes, the technologies suggested are:

- Cool white metal halide lamps.

- Cool white fluorescent lamps (inductive lamps offer a longer lifetime).

Figure 8.17. An example of a luminaire that focuses the emitted light on the space to be illuminated and reduces light pollution and glare.

Figure 8.18. Another example of a luminaire that focuses the emitted light and reduces light pollution and glare.

- LED systems can also be a good solution as they have a long lifetime and the appropriate color temperature can be selected, however, there are still some issues to consider such as cost, light flux, and efficacy. It is only a matter of time though before they will fulfill all the criteria and find a place in outdoor lighting (Figure 8.19). Another advantage of LEDs is that due to thermal management issues they perform better in cold environments, which is usually the case with outdoor spaces during the night.

- High-pressure xenon lamps can offer high light fluxes, good color rendering properties, and color temperatures that are required for a wide range of outdoor spaces, however, their efficacies and lifetime are lower than the above technologies.

Stadiums are a special case where high flux is required and good color rendering is also desired. Both peripheral and fovea vision must be supported so the best compromise is to use high-pressure xenon lamps or neutral white metal halides (Figure 8.20).

Figure 8.19. LED systems for outdoor lighting.

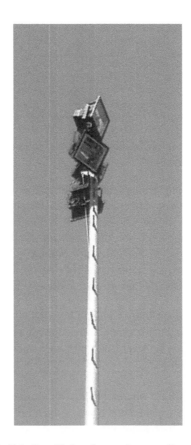

Figure 8.20. Metal halide floodlights for outdoor stadium lighting.

Figure 8.21. A tunnel illuminated with sodium lamps.

Although sodium lamps have dominated outdoor lighting for decades, the need for enhancing peripheral vision for security and safety reasons have made them replaceable in most outdoor applications. However, there are a few cases where they should continue to be used:

- Open roads or tunnels (Figure 8.21) where there are no pedestrians or significant traffic or where there are known obstacles, which cause safety concerns.

- Places where fog and dust are frequent due to the higher penetration of shorter wavelengths (yellow compared to blue).

- Heavy industry and warehouses where there is no safety/security issue or no dangerous work being undertaken.

- Near astronomical institutions as their longer than blue wavelengths scatter in the atmosphere less causing less light pollution.

- Airports and other facilities where high fluxes and good visibility is required but color rendering is not of importance.

An important issue to deal with when it comes to tunnels is the fact that one has to move from very bright environments to low light levels and vice versa in a very short time. This contrast causes adaptation difficulties for the drivers so either special lighting or specially built entrances and exits must be designed in order to prevent such abrupt changes (see also Section 1.2 in Chapter 1 on glare).

Figure 8.22. A halogen lamp with a motion detector.

With regard to security and safety lighting, when instant starting is required as, for example, in motion detecting systems (Figure 8.22) then halogen lamps are a good choice (with excellent color rendering properties) followed by LEDs.

When dealing with closed circuit television (CCTV), which is sensitive to IR radiation for nighttime operation, then the chosen light source technologies are halogen (with appropriate filters) and special IR emitting LEDs that can be found on the market for security applications.

The important thing about cameras is to keep all sources (natural and artificial) away from the field of view of the camera and also to ensure that the illumination is uniform so that shadows and dark spots are avoided.

It has also been discussed recently that blue lights deter drug users from using the area due to the difficulty in identifying their veins. Regardless of how effective this measure is, the lamp of choice in this case should be a blue LED system or a mercury fluorescent lamp with the appropriate phosphor.

8.3.3   Solar Lamps

When discussing outdoor lighting, it is also important to take into account the solar lamp systems that can be found on the market. A solar lamp is a light fixture equipped with a photovoltaic solar panel and a rechargeable battery that drives the lamp.

Solar lamps recharge during the day. When darkness sets at dusk, they turn on (usually automatically, although some of them include a switch for on, off, and

Figure 8.23. Solar lamps for garden lighting. The solar panels on top of the lamps recharge the batteries.

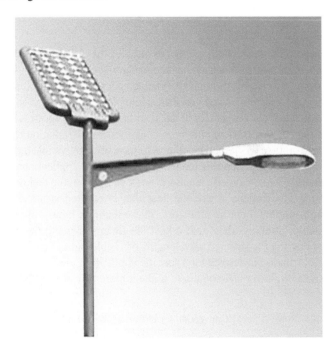

Figure 8.24. Solar lamp systems are also used for road lighting.

automatic), and remain illuminated overnight, depending on how much sunlight they receive during the day. Usually, the illumination/battery discharge lasts around 8 to 10 hours. In most solar lamp systems, the lamp of choice is LED technology.

Most solar lamps are used for outdoor lighting such as garden lighting (Figure 8.23) but when the solar panel is detached then the solar lamp can also be used in some indoor applications. Solar lamp systems for road lighting (Figure 8.24) can also be found on the market but such systems thus far have been used more widely in applications where not much light is required (gardens or footpaths) as they are easy to install and maintain, and provide a cheaper alternative to wired lamps.

The proposals discussed above show that each situation must be examined separately and the special requirements for every case must be evaluated. Different technologies offer different features and advantages so one must carefully consider the specific needs before choosing a technology and an illumination system.

8.4   Nonvisible Applications

Although the nonvisible emissions of the light sources discussed are not useful for lighting, there are some applications where they can be utilized. In fact, for several applications where nonvisible radiation is needed, light sources can even be modified by choosing the appropriate glass or phosphor in order to maximize their emissions in the desired wavelengths.

In this section, the focus is on the infrared (IR) and ultraviolet (UV) emissions of several lamps and the applications in which they are used.

8.4.1   Infrared Sources

When it comes to IR, the technology that dominates the discussion is the halogen lamp. Incandescent sources such as the halogen lamp emit a continuous spectrum extending from IR, passing through the visible spectrum, and reaching the UV. This continuous spectrum is what gives this technology its excellent color rendering index, which is ideal when color identification and color contrast is important.

The fact that most electrical energy is converted to IR radiation (heat) is, of course, a disadvantage when the objective is to have efficient visible light emitters but when heat is desired then this technology finds more applications. Table 8.1 shows the energy conversion efficiencies of various light source technologies.

For lighting applications where the heat that is emitted is not desired, the halogen lamps employ dichroic reflectors. These dichroic lamps, also known as *cold beam* lamps, employ filters that reflect the IR emissions to the back of the lamp while allowing the selected wavelengths of visible light to pass.

In case IR radiation is desired for a specific application, the lamp may employ special quartz with lower OH content to make it more transparent to the emitted IR waves.

Table 8.1. Power/Energy Conversion for Different *White* Light Sources

	Incandescent	Fluorescent	Metal Halide	LEDs
Visible light	10%	20%	30%	15%–25%
IR	70%	40%	15%	~ 0%
UV	0%	0%	20%	0%
Emitted energy	80%	60%	65%	15%–25%
Heat	20%	40%	35%	75%–85%
Total	100%	100%	100%	100%

Note: LEDs used for lighting applications do not emit IR red radiation as a byproduct but IR emitting LEDs exist for special applications.

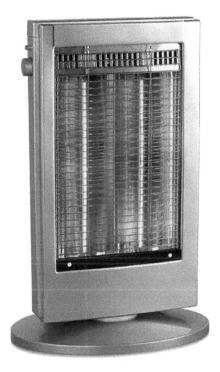

Figure 8.25. A home heater made of IR emitting tubes.

Some of those applications where heat is desired and halogen lamp technology is employed include:

- In bathrooms and saunas

- Food preparation areas in restaurants to keep food warm before serving

- For animal husbandry such as poultry brooding

- To provide warmth in places that are used for keeping young mammals such as zoos and veterinary clinics, or private areas where a range of animals are kept such as reptiles, amphibians, insects, and so on

- In house heating units (Figure 8.25)

- In some medical systems where dry heating is required

- In industrial manufacturing processes where high temperatures are required

Usually, incandescent lamps that are used for heating come with a red glass bulb. The amount of heat emitted is the same as it would be with a clear envelope. All the red filter does is cut off the light emitted at the other visible wavelengths.

8.4.2   Ultraviolet Sources

It is important to note on the subject of ultraviolet emissions from some of the lamps discussed because this is a potential danger to users. A lot of lamps emit ultraviolet radiation to different extents and although this is useful for certain applications it can also be dangerous in some cases.

Although ultraviolet radiation extends from 10 to 400 nm, we usually refer to three regions of the UV spectrum that are emitted by the sun.

Ultraviolet A, long wave	UV-A	400–315 nm
Ultraviolet B, medium wave	UV-B	315–280 nm
Ultraviolet C, short wave	UV-C	280–100 nm

The Earth's ozone layer blocks 97%–99% of this UV radiation from penetrating through the atmosphere and of the radiation that does get through, almost 99% is UV-A.

Ordinary soft glass (soda lime) is transparent to 90% of the light above 350 nm, but blocks over 90% of the light below 300 nm. Quartz glass, depending on quality, can be transparent to much shorter wavelengths (lower than 250 nm). Figure 8.26 shows the transparency curves of the two different kind of glass.

Fluorescent lamps and all lamps containing mercury emit ultraviolet radiation mainly at mercury's resonant emission lines at 254 and 185 nm but also a few emission lines in the UV-A (an emission line is a sharp and narrow atomic emission at a specific wavelength). The radiation at 185 and 254 nm is converted to visible emissions by the phosphor coated on the inside walls of the lamp but if one wants to make use of the radiation at 254 nm (which is the strongest line) wavelengths, then a mercury containing lamp made of quartz is used. The glass bulb blocks all radiation in the UV-B and UV-C but allows some emissions in

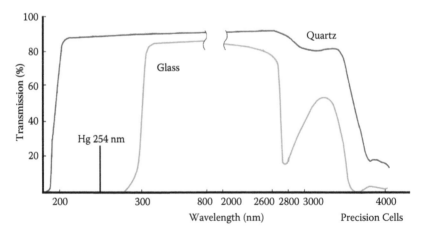

Figure 8.26. The spectral response of soft glass and quartz. It is clear from the diagram that mercury's strongest emission line at 254 nm can only go through quartz glass.

the UV-A to go through. This amount of UV-A is not harmful, but it is best to be avoided if the user is dealing with UV sensitive materials. According to the American National Electrical Manufacturers Association (NEMA), UV exposure from sitting indoors under fluorescent lights at typical office light levels for an 8-hour workday is equivalent to just over a minute of exposure to the sun. When the phosphor used converts those strong mercury emission lines not in visible but UV-A and/or UV-B radiation, then this kind of lamp can be used for various applications such as skin tanning, to attract insects, for decoration (black light), and so on.

Metal halide lamps usually have a second (outer) glass bulb to protect the inner components and prevent heat loss. This second glass envelope can also block most of the UV radiation emitted from mercury or other metals in the active medium. The cover glass of the luminaire is also used to block the UV, and can also protect people or equipment if the lamp should fail by exploding.

LEDs for general lighting emit narrow emission bands so it is one of the technologies that does not not pose any UV concerns at all, unless one deals with a diode specifically designed to emit in the UV for certain applications as those shown in Figure 8.27.

The continuous spectrum of incandescent lamps also extends into the UV region and attention must be paid to halogen lamps because their quartz glass is more transparent to this radiation. To avoid all risks, the user must be sure that the appropriate UV filters are used with halogen lamps.

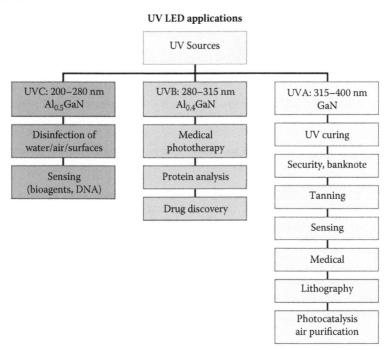

Figure 8.27. Ultraviolet emitting LEDs for various applications.

UV sources can be used for the following applications:

- 230–365 nm: UV-ID, label tracking, barcodes

- 240–280 nm: Disinfection, decontamination of surfaces, and water (DNA absorption has a peak at 260 nm)

- 250–300 nm: Forensic analysis, drug detection

- 270–300 nm: Protein analysis, DNA sequencing, drug discovery

- 280–400 nm: Medical imaging of cells

- 300–365 nm: Curing of polymers and printer inks

- 300–320 nm: Light therapy in medicine

- 350–370 nm: Bug zappers (flies are most attracted to light at 365 nm)

- Photochemotherapy: UV-A radiation in combination with psoralen is an effective treatment for psoriasis called *PUVA* (Figure 8.28); for reducing risks to the psoralen sensitive liver, this treatment is only used a limited number of times

- UV-B phototherapy is an effective long-term treatment for many skin conditions like vitiligo, eczema, and others

- UV radiation (mercury's 254 nm emissions) is used for sterilizing biology and medical-related spaces and wastewater treatment

- Ultraviolet radiation is used in several food processes to kill unwanted microorganisms

- UV fluorescent lamps are used in combination with heat-generating lamps in reptile housings

- UV-A and UV-B radiation is used for artificial skin tanning (sun beds/solarium); exposure to UV-B radiation leads to an increase of vitamin D but also of melanin in order to protect the skin from burns and genetic damage

Figure 8.28. A commercial discharge UV source for phototherapy applications such as psoriasis, eczema, and so forth.

For all intents and purposes, the user should always know whether the lamp being used is emitting in the UV region of the spectrum and what precautions have been taken.

References

1. Brown, G.M., Pandi-Perumal, S.R., Trakht, I., and Cardinali, D.P. (2009). Melatonin and its relevance to jet lag. *Travel Med. Infect. Dis.* 7 (2): 69–81.
2. Peased, Ken. (1999). A review of street lighting evaluations: Crime reduction effects. *Crime Prevention Studies* 10: 47–76. http://www.popcenter.org/library/crimeprevention/volume_10/.
3. Painter, Kate and Farrington, David P. (1999). Street lighting and crime: Diffusion of benefits in the Stoke-On-Trent project. *Crime Prevention Studies* 10: 77–122. http://www.popcenter.org/library/crimeprevention/volume_10/.
4. Clarke, Ronald V., Office of Community Oriented Policing Services (2008). Improving Street Lighting to Reduce Crime in Residential Areas. Problem-Oriented Guides for Police Response Guides Series Guide No. 8. http://cops.usdoj.gov/Publications/e1208-StreetLighting.pdf.
5. Atkins, S., Husain, S., and Storey, A. (1991). The Influence of Street Lighting on Crime and Fear of Crime. Crime Prevention Unit Paper 28. London, UK: Home Office.
6. Marchant, P.R. (2004). A demonstration that the claim that brighter lighting reduces crime is unfounded. *British Journal of Criminology* (3): 441–447.
7. He, Y., Rea, M.S., Bierman, A., and Bullough, J. (1997). Evaluating light source efficacy under mesopic conditions using reaction times. *J. Illumin. Eng. Soc.* 26: 125–38.
8. Lewis, A.L. (1999). Visual performance as a function of spectral power distribution of light sources at luminances used for general outdoor lighting. *J. Illumin. Eng. Soc.* 28: 37–42.

9 Light Sources and Health

9.1 Melatonin Suppression

In Chapter 1, the importance of the natural circadian rhythm and how it depends on levels of melatonin was discussed. In addition, we described how the use of warm and cool white light sources can simulate daylight changes in order to create more natural indoor environments and also how this dynamic lighting (Chapter 8) can create the appropriate lighting for a range of tasks and moods.

But one of the main objectives of circadian rhythm manipulation is to help groups of people at risk of facing serious consequences if this rhythm is disrupted. An example of how a disruption of the circadian rhythm can have on a person's health is the claim, supported by research, that illumination during sleep periods has dire effects on human health such as an increase in cancer cases. Breast cancer is the oncological condition whose relationship to circadian rhythm fluctuations has perhaps been studied most extensively [1–3]. Figure 9.1 shows melatonin suppression with respect to illuminance values.

People who wish to avoid such a disruption in their circadian rhythm must avoid blue light late in the evening when melatonin levels are supposed to increase. There are two solutions to this problem. People who wish to illuminate their space during the night should avoid light sources that emit in the blue region of the spectrum, so metal halides, cool white fluorescent, and LED lamps must not be used. Dimmed halogen lamps or warm white fluorescent and LED lamps (even yellow or red light LEDs) can offer the minimum illumination levels one might need during the night while also avoiding melatonin suppression. To be more specific, the most efficient melatonin-suppressing wavelengths fall within the range of 440–480 nm, although green light can also contribute to the suppression of

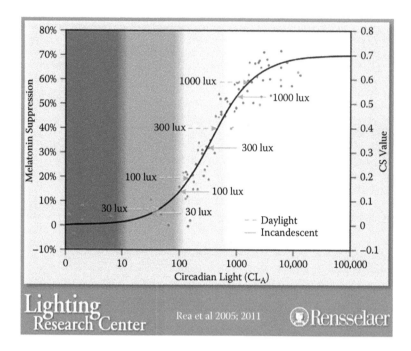

Figure 9.1. Melatonin suppression with different light sources and illuminance. (From Figueiro, M.G., Plitnick, B., Wood, B., and Rea, M.S., 2011, The Impact of Light from Computer Monitors on Melatonin Levels in College Students, *Neuro. Endocrinology Letters* 32: 158–163; Rea, M.S., Fiueiro, M.G., Bullough, J.D., and Bierman, A., 2005, A Model of Phototransduction by the Human Circadian System, *Brain Research Reviews* 50: 213–228.)

melatonin [4–5]. Light sources that emit in these ranges are fluorescent and LED lamps. These lamps can emit cool white light (rich in blue emissions) or even directly blue or green light with the use of appropriate phosphors or corresponding LEDs. Metal halide and xenon lamps can also emit cool white light but fluorescent and LED lamps offer greater availability and emission flexibility.

Another solution that has been suggested is the use of lenses that block blue light. Preventing melatonin deficiencies using lenses to block light of a low wavelength from reaching the retina presents a cost-effective, practical solution to the problem of increased malignancy rates in shift workers but also a broad range of conditions might respond to this inexpensive therapeutic tool such as common forms of insomnia, circadian rhythm disruption in shift workers, and perhaps even bipolar disorder [6].

On the other hand, if the concerns are the lack of focus and risk of accidents during night-shift work because the worker would be performing a task that requires alertness, then melatonin-suppressing short-wavelength light sources might be desired. Night-shift workers who wish to alter their circadian rhythms may use melatonin-suppressing light sources during the night and blue-filtered lenses during the day.

Other applications include the use of cool white light sources with significant blue emissions (fluorescent and LED lamps) to provide phototherapy treatment for people suffering from syndromes such as "delayed sleep phase syndrome" or "seasonal affective disorder." Bright light therapy for seasonal affective disorder (SAD) has been investigated and applied for over 20 years. Physicians and clinicians are increasingly confident that bright light therapy is a potent, specifically active, nonpharmaceutical treatment modality. Indeed, the domain of light treatment is moving beyond SAD and is now being used for nonseasonal depression (unipolar and bipolar), seasonal flare-ups of bulimia nervosa, circadian steep phase disorders, and more.

There are products, so-called *light boxes*, that are meant to be used as phototherapy devices for such cases. Figures 9.2 and 9.3 show such light therapy systems which can be found on the market.

People suffering from delayed sleep phase syndrome may get the recommended number of hours of sleep but they usually sleep late and wake up late compared to most. The use of light boxes can help them regulate and control their sleep patterns by manipulating their circadian rhythm.

For people with Alzheimer's disease, sleep disturbances are much more frequent and severe. Instead of consolidated sleep at night and wakefulness during the day, Alzheimer patients exhibit intermittent sleep throughout the 24-hour day. Poor sleep is one of the largest complaints among the elderly and studies have shown that when the elderly are exposed to high circadian light levels during the day and dim circadian levels at night, their sleep duration and efficiency improves significantly.

Seasonal affective disorder is another situation where people go through a period of depression after the summer months or in regions of the world with less daylight. Once again, light boxes (emitting daylight or cool white light) can help counterbalance this lack of light exposure.

There is still skepticism as to whether artificial sources reach the necessary illumination levels for the circadian rhythm to be affected and for such syndromes to be treated, but there is a lot of research and a number of publications that support this concept [7–13].

Figures 9.2. A commercial light therapy system.

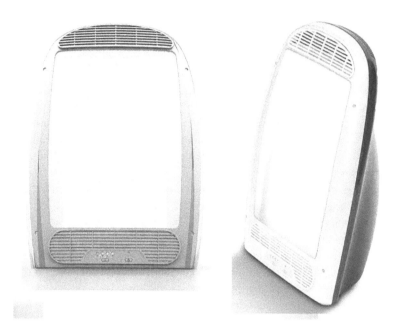

Figure 9.3. More examples of commercial light therapy systems.

Each light box reaches different illumination levels (from thousands of lux to a few hundred lux) depending on the source that is used and the duration of each phototherapy session. For example, the fact that boxes employing green light sources require lower illumination levels to reach the same result is supported by research [8].

If one requires the use of sources that emit at specific colors (narrow and near monochromatic emission bands), then fluorescent lamps with appropriate phosphors or specific color LEDs are the main options.

The important thing to remember is that in order to create a stimulating environment that enhances alertness or to treat various syndromes by affecting the circadian rhythm through melatonin suppression, one needs either high color temperature light sources such as metal halides, fluorescent, LED, and even xenon discharges (all with color temperatures >5000 K) or light with wavelengths shorter than 530 nm (blue and/or green light), so fluorescent and LEDs lamps should be used. In general, the use of fluorescent and LED lamps is proposed because they offer greater availability and flexibility in spectral output.

9.2 Light Therapy

In the previous section, we discussed the use of light sources in order to manipulate melatonin production, hence the circadian rhythm of humans. This manipulation takes place either through the use of multiple light sources used as dynamic systems in order to create different atmospheres and moods (different color temperatures) or with the use of short-wavelength emitting lamps to treat syndromes such as SAD and delayed sleep phase (blue-green sources or cool white sources).

In this section, we will mention a few more medical conditions that according to the literature can be treated through light therapy and phototherapy lamps. It is wise to advise the reader to consult a physician before a light therapy path is followed as this type of treatment, like any other, may require specific method-ologies and systems but also because the treatment might be damaging to people suffering from certain conditions that make them sensitive to light (for example, porphyria) or those who take photosensitive drugs.

The conditions treated by phototherapy can be broadly categorized into those that make use of short visible wavelengths (blue and/or green light) and those that use long visible wavelengths (red light). The following is a list of conditions according to this categorization:

- Short and medium visible wavelength light sources—Neonatal jaundice [14–16]

- Long visible wavelengths (red) light source—Further research is needed for these claims

  - Pain management

  - Wound healing

  - Hair growth

  - Blood treatment

  - Sinus-related diseases

- Mixed short and long visible wavelength light sources—Acne [17–20]

Phototherapy with mixed blue-red light, probably by combining antibacterial and anti-inflammatory action, is an effective means for treating mild-to-moderate acne vulgaris with no significant short-term adverse effects. Both fluorescent lamps emitting light of a specific color with appropriate phosphors and LED lamps of specific colors are used for these applications (Figure 9.4).

When small size light sources are required (to fit small commercial products or to treat small areas) and low levels of light are sufficient, then LED sources are chosen. For larger fluxes and when the size of the illuminating system is not an issue, then fluorescent lamps can also be used.

In addition to the use of fluorescent and LED lamps to treat the above condi-tions, another lamp that is frequently used in marketed products is the xenon flash lamp. Xenon flash lamps are used for *intense pulse light* treatments that include a list of applications such as hair removal, photo-rejuvenation (wrinkle removal), skin mark removal, and so on.

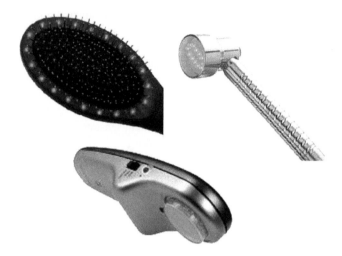

Figure 9.4. Commercial products employing red LED units that claim to have phototherapeutic results.

9.3   Flicker and Health Effects

Flicker is the modulation of a lamp's light output mainly caused by fluctuations of the mains voltage supply. The mains frequency is 60 Hz in the United States (50 Hz in Europe) causing a flicker at 120 Hz (100 Hz in Europe) in the lamp's output. Flicker is not perceptible at this frequency but it can still affect people. All types of lamps suffer from flicker (even LEDs that have a DC mode of operation) but people seemed to dislike, in particular, the flicker of fluorescent lamps operated with magnetic ballasts at that mains frequency. It is not uncommon to hear people in offices complain of headaches and dizziness due to the performance of these lamps [21,22]. In fact, research has shown that fluctuations of short-wavelength emissions are perceived to a higher extent [23,24].

For other groups of people who suffer from specific conditions, flicker can have more serious consequences. People suffering from photoepilepsy, for example, are most sensitive to flicker at 15 Hz, however a large percentage of them show signs of a photoconvulsive response at 50 Hz. People suffering from migraines are also more likely to be sensitive to lighting instabilities [25,26]. People with autism, too, could benefit from a solution to this problem. There are more conditions that are supposedly sensitive to flicker such as Ménière's disease, vertigo, Asperger's syndrome, Lupus, chronic fatigue syndrome, and Lyme disease, but regardless of how much these diseases correlate to flicker, everyone would like to see it minimized.

The way to suppress flicker is by changing the operating mode. For example, when electronic gear (30–50 kHz) replaced magnetic ballasts in fluorescent lamps, flicker was suppressed. Fluorescent lighting that operates from a magnetic control gear has a modulation of about 45% at a fundamental frequency of 100 Hz (in Europe). The same lamps that operate from an electronic control gear have a modulation of less than 15% at 100 Hz. Even DC-operated LED lamps flicker due to the fluctuation of the mains voltage. In all cases, the only flicker-free lamps available are those that employ self-correcting circuits (current feedback), although this type of design adds to the cost, which is why it is not used in all products. The best way for a consumer in Europe to test whether an LED lamp flickers or not (or check the degree of flicker in halogen and fluorescent lamps) is to look at the operating lamp through a digital camera display (such as a mobile phone camera display). Such a display usually refreshes at a rate of 30 frames per second and the flicker that is not visible with the naked eye becomes visible on the display. See Figures 9.5 and 9.6 for examples of flicker on photographs.

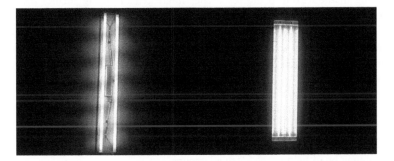

Figure 9.5. Ceiling lights as recorded by a mobile phone digital display (in Europe). The flicker is evident as stripes in the image for the magnetic ballast fluorescent lamps (left); while there is no flicker detection for the high-frequency electronic ballast fluorescent lamps (right).

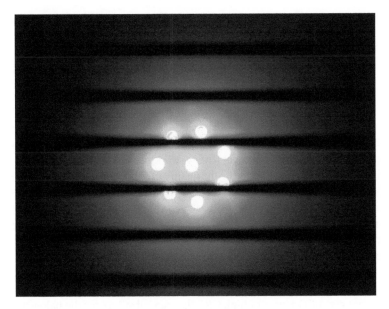

Figure 9.6. Image of an LED lamp. The stripes in the image reveal the flicker of the lamp.

Large fluctuations (i.e., large flicker percentages) are clearly recorded as striped images on the display. The distortions/stripes tend to fade with lower flicker percentages but they can still be detected by fine tuning the camera settings such as the exposure. Lamps that show no flicker or very small fluctuations (such as LED lamps with current feedback driving circuits) will be seen with no distortions on the display and the recorded images.

This easy and quick flicker test with such a widely available tool as a mobile phone may prove useful to consumers seeking to avoid the undesirable effects of flicker even at such nonvisible frequencies. It could also be useful as a quick screening method even to the researcher interested in taking further measurements of flickering lamps.

To summarize, the suggestion put forward here that relates to the problem of flicker is to use *warm white lamps* and preferably fluorescent lamps that employ electronic high-frequency gear and LED systems with self-correcting circuits (current feedback).

9.4   Light Sources and Eye Conditions

In this section, the focus is on eye conditions and diseases and how specific light sources may be of help to the affected groups.

Despite the extent to which these proposals can prove useful, the light sources discussed are the usual lamp technologies discussed throughout the book and used in many other applications.

Here are the basic rules based upon the proposals:

- Red color light penetrates more than shorter wavelengths and this might be important when dealing with loss of lens transmittance.

- Blue color light scatters more than longer wavelengths causing glare and discomfort.

- The peripheral vision (rods) is more sensitive to blue color light.

The choice of the right light source can help lighting engineers and doctors design the appropriate environment (living, working, and healing) for people with special visual needs, thus improving the quality of life for many groups of citizens.

As a general rule, older people benefit from light sources with intense red emissions (and warm white sources) because with age there is a reduction in the transmittance of light through the lens and increased scattering, particularly at short wavelengths. They also benefit from enhanced color contrast (high CRI sources) and higher illumination levels but with the avoidance of glare. The *halogen lamp* gives the highest color rendering and emits mainly in the red region so color contrast and light penetration through the lens is optimal. Warm white fluorescent and LED lamps of high CRI also can substitute for the halogen lamp.

However, the lighting engineer/designer should always take into account not only the age of the people living or working in the area that is to be lit but also any specific health conditions and how the selected light sources may affect (or not affect) the circadian rhythm already discussed. It must be stressed that people suffering from a condition should first seek medical attention to identify the specific cause and nature of their problems. Another important point is that people with impaired sight will always benefit from making their tasks and everyday life as simple as possible. People who do not suffer from any condition will not be affected by the proposals but hopefully even the slightest effect on the people in need will improve their quality of life.

The four most common causes of low vision in developed countries are cataracts, macular degeneration, glaucoma, and diabetic retinopathy [27,28]. These conditions involve different parts of the eye but in all cases the basic principles apply.

A *cataract* is a clouding that develops in the crystalline lens of the eye increasing the absorption and scatter of light. More light can help overcome the increased absorption but if scattering is high it will reduce visual capabilities (higher sensitivity to glare).

A good way to enhance color discrimination for older people without using high levels of illumination (to avoid glare) is simply to use a source with a high color rendering index.

Early in the development of an age-related cataract, the power of the lens may be increased, causing near-sightedness (myopia) and the gradual yellowing and opacification of the lens may reduce the perception of blue colors. A source with intense red emissions might help overcome the above issues as longer wavelengths will penetrate the lens and the rest of the eye to a higher degree, and will be refracted and scattered less.

Halogen lamps and lamps based on *sodium discharge* have little energy below 450 nm, so they should be preferred, but sodium discharge lamps have poor color rendering properties so should only be used when efficacy is important. *Fluorescent and LED lamps of warm white light tones and rich in red emissions* should also be used. Metal halide lamps have significant short-wavelength emissions so they should be avoided.

Another population that has a problem with exposure to short wavelengths are postoperative cataract patients who have had their lens removed. Such patients are much more likely to suffer photochemical retinal damage due to short-wavelength visible and UV radiation exposure than are people with their biological lens intact. Halogen lamps with UV filters and fluorescent or LEDs with increased red emissions should be the first choices.

This proposal goes along with the practice of elderly people benefiting from sunglasses that cut off most light at wavelengths below 550 nm, thus reducing eye lens fluorescence and scatter from the short visible wavelengths and UV radiation.

Glaucoma is an eye disorder in which the optic nerve suffers damage, permanently damaging vision in the affected eye(s) and progressing to visual field loss and complete blindness if untreated. It is often, but not always, associated with increased pressure of the fluid in the eye (aqueous humor). The treatment for glaucoma is based on reducing intraocular pressure, either with medication or surgery.

Since glaucoma first affects the peripheral vision and worsens the ability to see at night, light sources emitting light with medium and long-wavelength emissions (neutral and warm white sources) might prove useful. Foveal vision is characterized by a peak at 555 nm (green-yellow) of the eye sensitivity curve. Halogen and warm white or neutral white fluorescent and LED lamps are recommended. UV filters and minimization of blue emissions are desired in order to reduce fluorescence, glare, and further damage.

Age-related macular degeneration is a medical condition which results in a loss of vision in the center of the visual field (the macula) because of damage to the retina. The macula of the retina, which is a circular, yellow-pigmented area of the retina, 2–3 mm in diameter and centered on the fovea, changes in appearance. Macular degeneration can make it difficult to read or recognize faces, although enough peripheral vision remains to allow other activities in daily life. Higher levels of illumination (task lighting) and magnification of images can help at the early stages.

Because the peripheral vision outside the macula is unaffected one might consider the use of light sources with shorter wavelength emissions (blue or cool white lamps); however, the use of such sources may accelerate the damage due to the detrimental effect and photo-oxidative stress that short wavelengths have on the retina. The only proposal here is to use *high color rendering sources* (warm white) in order to increase color contrast and increase the illumination levels.

Diabetic retinopathy, as the name implies, is damage to various parts of the retina due to diabetes. The changes it produces in the vascular system that supplies the retina can have effects on visual capability depending on where on the retina the damage occurs. The simple rule of medium and long-wavelength sources for reduced peripheral vision applies here too.

Retinitis pigmentosa (RP) is a group of genetic eye conditions that leads to incurable blindness. In the progression of symptoms for RP, night blindness generally precedes tunnel vision by years or even decades. Since retinitis pigmentosa first affects night vision and then proceeds to tunnel vision, sources emitting medium and longer wavelengths might again prove useful. *Halogen lamps or neutral and warm white fluorescent and LED lamps* are proposed. Patients with retinitis pigmentosa can also have increased levels of intraocular light scatter so the choice of these light sources also helps in this respect.

The conditions discussed throughout this chapter cannot obviously be treated with the use of light sources alone. In fact, in most cases the light sources play more of a complementary and assisting role. Even when research supports the proposals made here, the subjects remain under scrutiny and are open to further examination and more research.

It also goes without saying that people suffering from any kind of condition mentioned in this book should first seek medical advice before deciding upon phototherapy as a path they will follow.

9.5   Color Vision Impairments

Care must be taken when color discrimination is important, as a percentage of people have some color vision deficiencies. The online encyclopedia

Wikipedia [29] provides a short description of all inherited or congenital color deficiencies. Color blindness may be acquired or inherited, with three types of inherited or congenital color vision deficiencies: monochromacy, dichromacy, and anomalous trichromacy.

Monochromacy, also known as *total color blindness,* is the lack of ability to distinguish colors, caused by the absence or defect of a cone. Monochromacy occurs when two or all three of the cone pigments are missing and color vision is reduced to one dimension.

- Rod monochromacy (achromatopsia) is an exceedingly rare, nonprogressive inability to distinguish any colors as a result of absent or nonfunctioning retinal cones. It is associated with light sensitivity (photophobia), involuntary eye oscillations (nystagmus), and poor vision. Rod monochromacy is the condition of having only rods in the retina. A rod monochromat is truly unable to see any color and can see only shades of gray.

- Cone monochromacy is the condition of having both rods and cones, but only a single kind of cone. A cone monochromat can have good pattern vision at normal daylight levels, but will not be able to distinguish hues. In humans, who have three different types of cones, there are three differing forms of cone monochromacy. There are three types named according to the single functioning cone class:

  1. *Blue cone monochromacy,* also known as *S-cone monochromacy*

  2. *Green cone monochromacy,* also known as *M-cone monochromacy*

  3. *Red cone monochromacy,* also known as *L-cone monochromacy*

Dichromacy is a moderately severe color vision defect in which one of the three basic color mechanisms is absent or not functioning. It is hereditary and, in the case of protanopia or deuteranopia, sex-linked, affecting predominantly males. Dichromacy occurs when one of the cone pigments is missing and color is reduced to two dimensions.

- Protanopia is a severe type of color vision deficiency caused by the complete absence of red retinal photoreceptors. It is a form of dichromatism in which red appears dark. It is hereditary, sex-linked, and present in 1% of males.

- Deuteranopia is a color vision deficiency in which the green retinal photoreceptors are absent, moderately affecting red-green hue discrimination. It is a form of dichromatism in which there are only two cone pigments present. It is likewise hereditary and sex-linked.

- Tritanopia is a very rare color vision disturbance in which there are only two cone pigments present and a total absence of blue retinal receptors.

Anomalous trichromacy is a common type of inherited color vision deficiency, occurring when one of the three cone pigments is altered in its spectral sensitivity. This results in an impairment, rather than loss, of trichromacy (normal three-dimensional color vision).

- Protanomaly is a mild color vision defect in which an altered spectral sensitivity of red retinal receptors (closer to green receptor response) results in poor red-green hue discrimination. It is hereditary, sex-linked, and present in 1% of males.

- Deuteranomaly, caused by a similar shift in the green retinal receptors, is by far the most common type of color vision deficiency, mildly affecting red-green hue discrimination in 5% of males. It is hereditary and sex-linked.

- Tritanomaly is a rare, hereditary color vision deficiency affecting blue-yellow hue discrimination. Unlike most other forms, it is not sex-linked.

Based on clinical appearance, color blindness may be described as total or partial. Total color blindness is much less common than partial color blindness. There are two major types of color blindness: those who have difficulty distinguishing between red and green, and those who have difficulty distinguishing between blue and yellow.

- Total color blindness
- Partial color blindness

Red-green:

- Dichromacy (protanopia and deuteranopia)
- Anomalous trichromacy (protanomaly and deuteranomaly)

Blue-yellow:

- Dichromacy (tritanopia)
- Anomalous trichromacy (tritanomaly)

Based on each individual case, the appropriate light sources should be used in order for the light emitted to be detected and discriminated from the others. LED systems offer a wide variety of colors to choose from and fluorescent lamps with appropriate phosphors can also offer color flexibility.

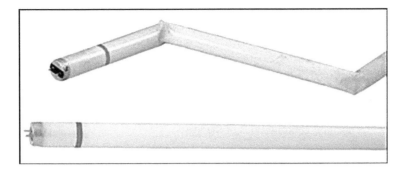

Figure 9.7. Coated fluorescent lamp for containment of mercury in case of breakage.

9.6 Mercury Issues

Mercury poisoning is a disease caused by exposure to mercury or its compounds. Toxic effects include damage to the brain, kidneys, and lungs. Mercury poisoning can result in several diseases, including acrodynia (pink disease), Hunter–Russell syndrome, and Minamata disease. Symptoms typically include sensory impairment (vision, hearing, speech), disturbed sensation, and a lack of coordination. The psychological symptoms associated with mercury poisoning are said by some to have inspired the phrase "mad as a hatter."

In the European Union, the directive on the Restriction of the Use of Certain Hazardous Substances in Electrical and Electronic Equipment (RoHS) bans mercury from certain electrical and electronic products, and limits the amount of mercury in other products.

Mercury can still be found in metal halide lamps and high-pressure sodium lamps (although most have steadily phased it out), but it is still the active medium in fluorescent lamps. Breakage of these lamps demands thorough cleaning and airing of the room. On the market, one can find specially (silicon) coated fluorescent lamps for material containment in case of breakage (Figure 9.7).

9.7 Photophobia

Photophobia is a symptom of an abnormal intolerance to light. When people suffer from photophobia, the cause may be an increased photosensitivity of the eyes or some other medical condition related to the eyes or nervous system [30].

Causes of photophobia that relate directly to the eye itself, include:

- Achromatopsia

- Aniridia

- Anticholinergic drugs may cause photophobia by paralyzing the iris sphincter muscle

- Aphakia (absence of the lens of the eye)

- Buphthalmos (abnormally narrow angle between the cornea and iris)

- Cataracts

- Cone dystrophy

- Congenital abnormalities of the eye

- Viral conjunctivitis ("pink eye")

- Corneal abrasion

- Corneal dystrophy

- Corneal ulcer

- Disruption of the corneal epithelium, such as that caused by a corneal foreign body or keratitis

- Ectopia lentis

- Endophthalmitis

- Eye trauma caused by disease, injury, or infection such as chalazion, episcleritis, glaucoma, keratoconus, or optic nerve hypoplasia

- Hydrophthalmos, or congenital glaucoma

- Iritis

- Optic neuritis

- Pigment dispersion syndrome

- Pupillary dilation (naturally or chemically induced)

- Retinal detachment

- Scarring of the cornea or sclera

- Uveitis

Neurological causes for photophobia include:

- Autism spectrum disorders

- Chiari malformation

- Dyslexia

- Encephalitis including myalgic encephalomyelitis also known as chronic fatigue syndrome

- Meningitis

- Subarachnoid hemorrhage

- Tumor of the posterior cranial fossa

Other causes include:

- Ankylosing spondylitis, which causes uveitis as an extra-articular feature

- Albinism

- Ariboflavinosis

- Benzodiazepines (long-term use of or withdrawal from benzodiazepines)

- Chemotherapy

- Chikungunya

- Cystinosis

- Ehlers–Danlos syndrome

- Hangover

- Influenza

- Infectious mononucleosis

- Magnesium deficiency

- Mercury poisoning

- Migraine

- Rabies

- Tyrosinemia type II, also known as Richner–Hanhart syndrome

Several symptoms such as fatigue, headaches, stress, hypertension, and others may also be the result of over-illumination of a space.

If someone suffers from photophobia, then medical attention should be sought in order to find the cause. To improve the lighting environment for people who are uncomfortable with certain levels of illumination, dimming/controlled lighting systems should be chosen. Halogen and LED systems can be easily dimmed to any level.

9.8   UV Exposure

As a large number of lamps (fluorescent, metal halide, halogen) also emit in the UV range, care must be taken so that people are shielded from this type of radiation using filters or protective covers. Generally speaking, there is no danger from fluorescent lamps when the exposure distance is more than a few tens of centimeters from the source (compared to the exposure one would get from sunlight).

However, there is a long list of conditions from which people suffer and anecdotal or official studies show that UV radiation from light sources such as fluorescent lamps can have negative effects and enhance the symptoms of the sufferers [31,32].

This list includes the following:

- Migraines

- Solar urticaria (allergy to sunlight)

- Systemic lupus erythematosus

- Photophobia

- Patients undergoing photodynamic therapy

- Chronic fatigue syndrome

- Phytophotodermatitis

- Actinic prurigo

- Lupus

- Chronic actinic dermatitis

- Squamous cell carcinoma

- Taking certain phototoxic drugs

- Polymorphous light eruption

- Psoriasis

- Atopic eczema

- Erythema multiforme

- Seborrheic dermatitis

- Immunobullous diseases

- Mycosis fungoides

- Smith–Lemli–Opitz syndrome

- Porphyria cutanea tarda

- Sjögren's syndrome

- Senear–Usher syndrome

- Rosacea

- Dermatomyositis

- Darier's disease

- Kindler–Weary syndrome

In all cases, care must be taken to shield people from UV radiation with filters and protective covers or the use of LED lighting that has no emission byproducts in the UV or IR range (unless they are UV or IR LEDs). Table 9.1 shows some of the photobiological hazards of skin and eye exposure to different spectral regions.

All the proposals made here are based on the properties of light discussed in this book and the technological and spectral characteristics of the available light sources. As a lighting guide these proposals may offer solutions to several health-related conditions but they also fit well within the normal practice of light source selection and, in particular, home lighting where warm white and high color rendering sources are usually preferred.

Table 9.1. Photobiological Hazards

Hazard	Wavelength Range (nm)	Principle Bioeffects	
		Skin	Eye
Actinic UV	200–400	Erythema, Elastosis	Photokeratitis, Cataractogenesis
Near UV	315–400	—	Cataractogenesis
Retinal blue light	300–700	—	Photoretinitis
Retinal thermal	380–1400	—	Retinal burn
Infrared radiation	780–3000	—	Corneal burn, Cataractogenesis
Thermal	380–3000	Skin burn	—

References

1. Schernhammer, E.S. and Schulmeister, K. (2004). Melatonin and cancer risk: Does light at night compromise physiologic cancer protection by lowering serum melatonin levels? *British Journal of Cancer* 90 (5): 941–943.
2. Hansen, J. (2001). Increased breast cancer risk among women who work predominantly at night. *Epidemiology* 12 (1): 74–77.
3. Davis, S., Mirick, D.K., and Stevens, R.G. (2001). Night-shift work, light at night, and risk of breast cancer. *Journal of the National Cancer Institute* 93 (20): 1557–1562.
4. Wright, H.R. and Lack, L.C. (2001). Effect of light wavelength on suppression and phase delay of the melatonin rhythm. *Chronobiology International* 18 (5): 801–808.
5. Wright, H.R., Lack, L.C., and Partridge, K.J. (2001). Light-emitting diodes can be used to phase delay the melatonin rhythm. *Journal of Pineal Research* 31 (4): 350–355.
6. Kayumov, L. (2005). Blocking low-wavelength light prevents nocturnal melatonin suppression with no adverse effect on performance during simulated shift work. *Journal of Clinical Endocrinology & Metabolism* 90 (5): 2755–2761.
7. Wright, H.R., Lack, L.C., and Kennaway, D.J. (2004). Differential effects of light wavelength in phase advancing the melatonin rhythm. *Journal of Pineal Research* 36 (2): 140–144.
8. Ahmed, S., Cutter, N.L., Lewy, A.J., Bauer, V.K., Sack, R.L., and Cardoza, M.S. (1995). Phase response curve of low-intensity green light in winter depressives. *Sleep Research* 24: 508.
9. Strong, R.E., Marchant, B.K., Reimherr, F.W., Williams, E., Soni, P., and Mestas, R. (2009). Narrow-band blue-light treatment of seasonal affective disorder in adults and the influence of additional nonseasonal symptoms. *Depress. Anxiety* 26 (3): 273–278.
10. Paul, Michel A., Miller, James C., Gray, Gary, Buick, Fred, Blazeski, Sofi, and Arendt, Josephine. (2007). Circadian phase delay induced by phototherapeutic devices. *Sleep Research* 78 (7): 645–652.
11. Gooley, J.J., Rajaratnam, S.M.W., Brainard, G.C., Kronauer, R.E., Czeisler, C.A., and Lockley, S.W. (2010). Spectral responses of the human circadian system depend on the irradiance and duration of exposure to light. *Science Translational Medicine* 2 May (31): 31–33.
12. Thompson, C., Stinson, D., and Smith, A. (1990). Seasonal affective disorder and season-dependent abnormalities of melatonin suppression by light. *Lancet* 336 (8717): 703–706.
13. Smith, M.R. and Eastman, C.I. (2008). Night-shift performance is improved by a compromise circadian phase position: Study 3, Circadian phase after 7 night shifts with an intervening weekend off. *Sleep* 31 (12): 1639–1645.
14. Ennever, J.F., Sobel, M., McDonagh, A.F., and Speck, W.T. (1984). Phototherapy for neonatal jaundice: *In vitro* comparison of light sources. *Pediatr. Res.* 18 (7): 667–670.

15. Stokowski, L.A. (2006). Fundamentals of phototherapy for neonatal jaundice. *Adv Neonatal Care* 6 (6): 303–312.
16. Amato, M. and Inaebnit, D. (1991). Clinical usefulness of high-intensity green light phototherapy in the treatment of neonatal jaundice. *Eur. J. Pediatr.* 150 (4): 274–276.
17. Kawada, A., Aragane, Y., Kameyama, H., Sangen, Y., and Tezuka, T. (2002). Acne phototherapy with a high-intensity, enhanced, narrow-band, blue light source: An open study and *in vitro* investigation. *J. Dermatol. Sci.* 30 (2): 129–135.
18. Papageorgiou, P., Katsambas, A., and Chu, A. (2000). Phototherapy with blue (415 nm) and red (660 nm) light in the treatment of acne vulgaris. *Br. J. Dermatol.* 142 (5): 973–978.
19. Goldberg, D.J. and Russell, B.A. (2006). Combination blue (415 nm) and red (633 nm) LED phototherapy in the treatment of mild to severe acne vulgaris. *J. Cosmet. Laser Ther.* 8 (2): 71–75.
20. Goldberg, D.J., Amin, S., Russell, B.A., Phelps, R., Kellett, N., and Reilly, L.A. (2006). Combined 633-nm and 830-nm LED treatment of photoaging skin. *J. Drugs Dermatol.* 5 (8): 748–753.
21. Berman, S.M., Greenhouse, D.S., Bailey, I.L., Clear, R., and Raasch, T.W. (1990). Human electroretinogram responses to video displays, fluorescent lighting and other high-frequency sources. *Optometry and Vision Science* 68 (8): 645–662.
22. Wilkins, A.J., Nimmo-Smith, I.M., Slater, A., and Bedocs, L. (1989). Fluorescent lighting, headaches and eye-strain. *Lighting Research and Technology* 21 (1): 11–18.
23. Wilkins, A.J. and Clark, C. (1990). Modulation from fluorescent lamps. *Lighting Research and Technology* 22 (2): 103–109.
24. Wilkins, A.J. and Wilkinson, P. (1991). A tint to reduce eye-strain from fluorescent lighting? Preliminary observations. *Ophthalmic and Physiological Optics* 11: 172–175.
25. Jeavons, P.M. and Harding, G.F.A. (1972). *Photosensitive Epilepsy: A Review of the Literature and a Study of 460 Patients.* London: William Heinemann Books.
26. Good, P.A., Taylor, R.H., and Mortimer, M.J. (1991). The use of tinted glasses in childhood migraine. *Headache* 31: 533–536.
27. All about Vision. http://www.allaboutvision.com.
28. World Health Organization. Blindness. http://www.who.int/topics/blindness/en.
29. Wikipedia. Color Blindness. http://en.wikipedia.org/wiki/Color_blindness.
30. Wikipeida. Photophobia. http://en.wikipedia.org/wiki/Photophobia.
31. Scientific Committee on Emerging and Newly Identified Health Risks SCENIHR, Light Sensitivity. http://ec.europa.eu/health/ph_risk/committees/04_scenihr/docs/scenihr_o_019.pdf.
32. European Guidelines for Photodermatoses. http://www.euroderm.org/edf/images/stories/guidelines/guideline_Photoaggravated_dermatoses.pdf.

Epilogue

One can acknowledge the attractive features of incandescent lamps, such as the high color rendering index of the white light they emit and therefore the near perfect color reproduction of the illuminated objects or the robustness and long life of the light-emitting diodes or the efficacy of the electrical discharges. All technologies have attractive characteristics and can still play a role in different areas. If that was not the case and a light source product or technology could not offer unique or competitive features, then it probably would have been obsolete, retired, and most likely mentioned only as a historical footnote next to gas and oil lamps. When one selects a light source to be used in a specific area, many parameters must be taken into account, such as the tasks performed in that area, the age of the people who occupy the area, the illuminance levels required, any medical conditions of the occupants, the overall cost that might or might not be attainable, and so on. However, once the principles on which the proposals are based are taken into account, then any situation can be evaluated and the right selection of light sources can be made.

If one considers what an ideal artificial light source would be like, then it becomes quite apparent that it will be difficult to develop it and it certainly does not exist today. For that reason, all or at least most of the technologies discussed in this book will continue to serve us unless a revolution in the light source research field takes place.

But what would an ideal light source be like? What would be its characteristics? Let us look at a list of features and the above point will emerge as each technology examined meets some but not all of them.

- *High luminous efficiency*—Currently, white sources reach 120 lm/W while the monochromatic low-pressure sodium lamp reaches 200 lm/w. A source that exceeds these limits is desirable and will have a high impact from an energetic and economic aspect.

- *Long lifetime*—Induction lamps (low-pressure mercury and high-pressure sulfur) exceed 20,000 hours while LEDs exceed 50,000 hours. A technology that consistently exceeds 150,000 hours is a realistic target.

- *Near perfect color rendering*—Incandescent lamps are assigned a value of 100 while several phosphors used with some of the other technologies can approach this.

- *Color temperatures*—Warm white is preferred by the majority of consumers but the possibility to control the tones in a dynamic system would be ideal.

- *Avoidance of flickering* and of electromagnetic interference with other electronic equipment.

- *Attractive shape and form*—Compact, light, strong, attractive, and exchangeable with other types of lamps and fitting existing infrastructure (manufacturing and power supplies).

- *Environmentally friendly*—The avoidance of toxic, harmful, or rare materials (such as mercury in fluorescent lamps or indium in LEDs) as well as recyclable.

- *Useful emitter*—The minimization of wasted energy in the form of heat or UV unless a special application requires it.

- *Low cost*—Cheap for the consumer but profitable for the industry.

For the time being, no light source can satisfy all of these features and it is quite a challenge to develop one that does. At the moment, the improvement of one characteristic might lead to the compromise of another so one can only use the products depending on priorities and applications. The world and the industry would definitely change with such an ideal technology.

But are there any changes taking place now? And what are the most pressing matters for the related researchers and developers?

Without a doubt, mercury plays a central role in the market of light sources as an element that is found in the majority of discharge lamps from low-pressure mercury fluorescent lamps to high-pressure high-intensity discharges (HIDs) to super-high-pressure projection lamps. Even in high-pressure sodium lamps and high-pressure metal halide lamps, mercury is usually added and acts as one of the main active media. Light source scientists and engineers today are asked to not only push the upper limit of light source efficiency to higher values but also to eradicate mercury and search for new material that will serve as efficient radiators. The replacement of mercury in lighting products is a requirement that stems from a long-lasting environmental concern toward mercury, particularly its organic form that is hazardous and life threatening, and has found a new frame these days in the form of government official directives (the well-known RoHS directives where the acronym stands for Reduction of Hazardous Substances). Other than mercury, the use of lead, cadmium, and hexavalent chromium is also not allowed. The immediate need to replace mercury for environmental reasons will probably have some drastic effects on the way low-pressure discharge lamps, such as the fluorescent lamp, are developed. The molecular low-pressure discharge lamps described in this book could prove to be a dynamic new protagonist.

The target for the efficiency and/or efficacy is to reach the same value or higher than what is achieved with the use of current products (such as the fluorescent lamps that can reach 120 lm/W although there is usually a trade-off between efficacy and color rendering index depending on the chosen phosphor for different applications). Lamps based on the principle of discharges in high-pressure vapors already show high luminous efficacies and a wide range of color temperatures and rendering indices, so they will play an important role in the decades ahead.

It can be said that LED technology is the least researched technology, as it has been the latest one to penetrate the market. Most of the changes and areas of research in LED technology are discussed in detail in Chapter 4. From the inner mechanisms of the crystals to the way they are packaged and manufactured are all areas where improvements will probably take place soon. So in the near future, lighting will be defined by the developments in solid-state lighting and the rapid increases in LED efficacies.

Unless a developing (organic light-emitting diodes seem promising but their progress is still in the early stages) or a new technology overtakes the existing ones, it seems that LEDs will penetrate and dominate an ever-increasing number of applications. As the cost of LEDs decreases and their efficacies and flux increase, LEDs become the dominant force of the lighting market. Their main advantages are the wide range of spectral outputs they offer, the dynamic lighting capabilities that certain multiple unit LED systems have, their relatively low-power consumption, their flexible geometries, and their lifetimes. From working spaces to outdoor illumination and from ambient lighting to architectural effects, LED systems seem to be able to accommodate pretty much all needs. Indeed, if one envisions a lighting system that could be installed and by appropriate programming accommodate any need or application imagined, then LED technology seems to be the closest one to realizing such a scenario. The market and all lighting and light source-related events testify to this as this solid-state technology already plays a main role in every single case.

The technology of quantum dots, new advanced phosphors, pulsing operation, and other methods for converting and manipulating spectra of existing light sources will also offer new solutions if developed and might lead the whole scientific community to reexamine and rethink all existing technologies, as their efficacies can benefit from such developments.

Many of the light sources mentioned are the result of combining different existing technologies. Starting with high-pressure mercury discharges, different other products were developed including metal halide lamps. The inductive operation principle of mercury fluorescent lamps was applied to sulfur lamps. In high-pressure xenon or sodium lamps mercury has been added. The hybrid mercury lamp is a combination of the incandescent principle and mercury high-pressure discharge. The polycrystalline alumina burner (PCA) that was first developed for sodium lamps found use in the development of the metal halide family of lamps. All these examples show that although the field of research and development of light sources is waiting for a new revolution in ways to create light, until that revolution takes place, the field will be dominated by combinations of existing technologies that will satisfy new needs or applications, and the optimization or improvement of existing products.

Lighting and light sources represent more than just cheap electric energy-based tools to counterbalance the absence of light at nighttime or even a pretty serious multibillion dollar market. From the improvement of the quality of life of people in their working and living spaces, to road safety or treatment of medical conditions, and from managing cancer rates to improving the lives of vision-impaired people, the correct choice of light sources and the appropriate lighting design can indeed make the difference between comfortable and problematic everyday life or indeed life and death. Time will tell if there will be a clear winner in this light sources race for dominance, but for now, every technology and lamp product still offers unique features and advantages, and the goal of this book was to help the reader match those attributes to different situations and needs.

Glossary

Sources: GE Lighting (www.gelighting.com/LightingWeb/emea/resources/world-of-ge-lighting/glossary/) *and Philips Lighting* (www.lighting.philips.com/main/connect/lighting_university/glossary/) *Web Sites.*

Accent Lighting: Directional lighting used to emphasize a particular object or draw attention to a display item.

Ambient Lighting: The general lighting present in an area that is set to create an atmosphere. It is a combination of general, architectural, and task lighting.

Arc Tube: A sealed quartz or ceramic tube where the electrical discharge (arc) occurs and generates light.

Architectural Lighting: Lighting that is meant to highlight the building and space itself (walls, ceilings, floors, etc.) rather than the objects inside the space.

Average Rated Lifetime: This is the time duration, beyond which, from an initially large number of lamps under the same construction and under controlled conditions, only 50% still function. Measurements of rated average lamp lifetimes are usually made by applying an operating cycle. For example, the lamps can be operated for 18 hours a day and remain switched off for the other 6 hours, or 3 hours on and 1 hour off. Such measurements offer a good basis for comparisons on technical life and reliability, although the same figures are unlikely to be obtained in practice, because parameters such as supply voltage, operating temperature, absence of vibration, switching cycle, et cetera, will always be different.

Ballast: An auxiliary piece of equipment required to start and to properly control the flow of current to gas discharge light sources such as fluorescent and high-intensity discharge (HID) lamps. Typically, magnetic ballasts (also called *electromagnetic ballasts*) contain copper windings on an iron core while electronic ballasts are smaller and more efficient and contain electronic components.

Base or Socket: The socket is the receptacle connected to the electrical supply; the base is the end of the lamp that fits into the socket. There are many types of bases used in lamps, screw bases being the most common for incandescent and HID lamps, while bipin bases are common for linear fluorescent lamps.

Bayonet: A style of bulb base that uses keyways instead of threads to connect the bulb to the fixture base. The bulb is locked in place by pushing it down and turning it clockwise.

Black Body: A hot body with an incandescent black surface at a certain temperature used as a standard for comparison. Note that a black surface is the best radiator possible. A tungsten filament will emit slightly less radiation than a black body at the same temperature.

Black Light: A popular term referring to a light source emitting mostly near-UV (320 to 400 nm) and very little visible light.

Burning Position: The position in which a lamp is designed to operate for maximum performance and safety.

Candela (cd): The measure of luminous intensity of a source in a given direction. The term has been retained from the early days of lighting when a standard candle of a fixed size and composition was defined as producing one candela in every direction.

Cathode: Metal filaments that emit electrons in a fluorescent lamp. Electrons emitted by the cathode are attracted to the positive electrode (anode), creating an electric current between the electrodes.

Ceramic Metal Halide: A type of metal halide lamp that uses a ceramic material for the arc tube instead of glass quartz, resulting in better color rendering (>80 CRI) and improved lumen maintenance.

Chromaticity: Measurement to identify the color of a light source, typically expressed as (x,y) coordinates on a chromaticity chart.

CIE: Abbreviated as CIE from its French title, Commission Internationale de l'Eclairage, the International Commission on Illumination is a technical, scientific, and cultural organization devoted to international cooperation and exchange of information among its member countries on matters relating to the science and art of lighting.

Color Rendering Index (CRI): An international system used to rate a lamp's ability to render object colors. The higher the CRI (based upon a 0–100 scale) the richer colors generally appear. CRI ratings of various lamps may be compared, but a numerical comparison is only valid if the lamps are close in color temperature.

Color Stability: The ability of a lamp or light source to maintain its color rendering and color appearance properties over its life. The color properties of some discharge light sources may tend to shift over the life of the lamp.

Color Temperature: Measured in Kelvin, CCT represents the temperature an incandescent object (like a filament) must reach to mimic the color of the lamp. Yellowish-white (*warm*) sources, like incandescent lamps, have lower color temperatures in the 2700–3000 K range; white and bluish-white (*cool*) sources have higher color temperatures.

Compact Fluorescent Lamp (CFL): The general term applied to fluorescent lamps that are single-ended and that have smaller diameter tubes that are bent to form a compact shape. Some CFLs have integral ballasts and medium or candelabra screw bases for easy replacement of incandescent lamps.

Contrast: The ratio of the luminance of an object to that of its immediate background.

Cool White: A lamp with a color temperature of 5000 to 6500 K.

Correlated Color Temperature (CCT): A term used for discharge lamps, where no hot filament is involved, to indicate that the light appears "as if" the discharge lamp is operating at a given color temperature. CCT generally measures the *warmth* or *coolness* of light source appearance using the Kelvin (K) temperature scale.

Cutoff Angle: The critical viewing angle beyond which a source can no longer be seen because of an obstruction (such as a baffle or overhang).

Cutoff Luminaire: IESNA classification that describes a luminaire having a light distribution in which the candela per 1000 lamp lumens does not numerically exceed 25 (2.5%) at or above an angle of 90° above nadir, and 100 (10%) at or above a vertical angle of 80° above nadir. This applies to all lateral angles around the luminaire.

Dichroic Reflector (or Filter): A reflector (or filter) that reflects one region of the spectrum while allowing the other region(s) to pass through. A reflector lamp with a dichroic reflector will have a "cool beam," that is, most of the heat has been removed from the beam by allowing it to pass through the reflector while the light has been reflected.

Dimmer, Dimming Control: A device used to lower the light output of a source, usually by reducing the wattage it is being operated at. Dimming controls are increasing in popularity as energy-conserving devices.

Disability Glare: Glare resulting in reduced visual performance and visibility.

Efficacy: A measurement of how effective the light source is in converting electrical energy to lumens of visible light. Expressed in lumens-per-watt this measure gives more weight to the green region of the spectrum and less weight to the blue and red region where the eye is not as sensitive.

Efficiency: The efficiency of a light source is simply the fraction of electrical energy converted to light, that is, watts of visible light produced for each watt of electrical power with no concern about the wavelength where the energy is being radiated. For example, a 100-watt incandescent lamp converts less than 10% of the electrical energy into light; discharge lamps convert more than 25% into light. The efficiency of a luminaire or fixture is the percentage of the lamp lumens that actually comes out of the fixture.

Electrical Discharge: A condition under which a gas becomes electrically conducting and becomes capable of transmitting current, usually accompanied by the emission of visible and other radiation. An electric spark in air is an example of an electrical discharge, as is a welder's arc and a lightning bolt.

Electrode: Any metal terminal emitting or collecting charged particles, typically inside the chamber of a gas discharge lamp. In a fluorescent lamp, the electrodes are typically metal filaments coated with special electron–emissive powders.

Electrodeless Lamps: Light sources where the discharge occurs in a chamber with no electrodes (no metal). The energy for the discharge is supplied by radio-frequency excitation, for example, microwaves.

Electromagnetic Ballast (Magnetic Ballast): A ballast used with discharge lamps that consists primarily of transformer-like copper windings on a steel or iron core.

Electronic Ballast: A short name for a fluorescent high-frequency electronic ballast. Electronic ballasts use solid-state electronic components and typically operate fluorescent lamps at frequencies in the range of 25–35 kHz. The benefits are: increased lamp efficacy, reduced ballast losses, and lighter, smaller ballasts compared to electromagnetic ballasts. Electronic ballasts may also be used with HID (high-intensity discharge) lamps.

Eye Sensitivity: The curve depicting the sensitivity of the human eye as a function of wavelength (or color). The peak of human eye sensitivity is in the yellow-green region of the spectrum. The normal curve refers to photopic vision or the response of the cones.

Filament: Metal tungsten wire heated by the passage of electrical current, used to emit light in incandescent lamps. In fluorescent lamps the filament is coated with emission mix and emits electrons when heated.

Flicker: The periodic variation in light level caused by AC operation that can lead to strobe effects.

Flood: Used to refer to the beam pattern of a reflector lamp, which disperses the light over a wide beam angle, typically 20 degrees or more. (*Flood* as opposed to *spot*.)

Floodlight: A luminaire used to light a scene or object to a level much brighter than its surroundings. Usually floodlights can be aimed at the object or area of interest.

Fluorescent Lamp: A high-efficiency lamp utilizing an electric discharge through inert gas and low-pressure mercury vapor to produce ultraviolet (UV) energy. The UV excites phosphor materials applied as a thin layer on the inside of a glass tube, which makes up the structure of the lamp. The phosphors transform the UV to visible light.

Footcandles: Unit of illuminance (light falling on a surface). One lumen falling on one square foot equals one-foot candle. Also, it is a measurement of the amount of light reaching a subject.

Fovea, Foveal Vision: A small region of the retina corresponding to what an observer is looking straight at. This region is populated almost entirely with cones, while the peripheral region has increasing numbers of rods. Cones have a sensitivity peaking in the green and correspond to the eye response curve.

Glare: Visual discomfort caused by excessive brightness is called *discomfort glare*. If task performance is affected it is called *disability glare*. Glare can be direct glare or indirect (reflected) glare.

Halogen Lamp: A halogen lamp is an incandescent lamp with a filament that is surrounded by halogen gases, such as iodine or bromine. Halogen gases allow the filaments to be operated at higher temperatures and higher efficacies. The halogen participates in a tungsten transport cycle, returning tungsten to the filament and prolonging lamp life.

Halophosphates: The class of phosphors commonly used in fluorescent lamps. Halophosphates are limited in their ability to provide a high color rendering index without sacrificing light output.

High-Bay Lighting: Lighting designed for (typically) industrial locations with a ceiling height of 7 meters and above.

High-Intensity Discharge (HID) Lamp: A general term for mercury, metal halide, and high-pressure sodium lamps. HID lamps contain compact arc tubes, which enclose mercury and various gases with other chemicals and operate at relatively high pressures and temperatures.

High-Pressure Sodium (HPS) Lamp: HPS lamps are high-intensity discharge light sources that produce light by an electrical discharge through sodium vapor operating at relatively high pressures and temperatures.

Hot Restart Time: The time it takes for a high-intensity discharge lamp to reach 90% of light output after going from on to off to on.

Illuminance: The *density* of light (lumens/area) incident on a surface; that is, the light level on a surface. Illuminance is measured in footcandles or lux.

Incandescent Lamp: A light source that generates light utilizing a thin filament wire (usually of tungsten) heated to white by an electric current passing through it.

Induction Lighting: Gases can be excited directly by radiofrequency or microwaves from a coil that creates induced electromagnetic fields. This is called *induction lighting* and it differs from a conventional discharge, which uses electrodes to carry current into the arc. Induction lamps have no electrodes inside the chamber and generally, therefore, have longer life than standard lamps.

Infrared Radiation: Electromagnetic energy radiated in the wavelength range of about 770 to 1,000,000 nanometers. Energy in this range cannot be seen by the human eye, but can be sensed as heat by the skin.

Instant Start Lamp: A fluorescent lamp, usually with a single pin at each end, approved to operate on instant start ballasts. The lamp is ignited by a high voltage without any filament heating.

Inverse Square Law: Formula stating that if you double the distance from the light source, the light level goes down by a factor of 4, if you triple the distance, it goes down by a factor of 9, and so on.

Isolux Plot (or Isofootcandle Plot): A line plotted to show points of equal illuminance (lux or footcandles) on a surface illuminated by a source or sources.

Kelvin: A unit of temperature starting from absolute zero. Zero Celsius (or Centigrade) is 273 K.

Kilowatt Hour: KWH. A unit of measurement for electrical energy. One kilowatt hour equals 1000 watts of energy used for 1 hour.

Lamp Types: Filament Lamps: Incandescent, halogen. Discharge Lamps: Fluorescent, HID (high-intensity discharge). HID Lamps: Mercury, HPS (high-pressure sodium), MH (metal halide), and CMH (ceramic metal halide).

Lens: A transparent or semitransparent element, which controls the distribution of light by redirecting individual rays. Luminaires often have lenses in addition to reflectors.

Light Tubes or Light Pipes: Used for transporting or distributing natural or artificial light. In their application to daylighting, they are also often called *sun pipes, solar light pipes,* or *daylight pipes.* Generally speaking, a light pipe or light tube may refer to a tube or pipe for transport of light to another location, minimizing the loss of light, or a transparent tube or pipe for distribution of light over its length, either for equidistribution along the entire length (sulfur lamps employed such tubes) or for controlled light leakage.

Low Voltage Lamps: Incandescent lamps that operate at 6, 12, or 24 volts. Low voltage lamps require a step-down transformer to reduce the voltage from the normal household 220 volts in Europe and 120 volts in the United States.

Lumens: The measure of the luminous flux or quantity of light emitted by a source. For example, a candle provides about 12 lumens. A 60-watt incandescent lamp provides about 840 lumens.

Luminaire: A complete lighting unit consisting of a lamp (or lamps) and ballast (or ballasts) as required together with the parts designed to distribute the light, position, and protect the lamps and connect them to the power supply. A luminaire is often referred to as a *fixture.*

Luminaire Efficiency: The ratio of total lumens emitted by a luminaire to those emitted by the lamp or lamps used in that luminaire.

Luminance: The measure of "surface brightness" when an observer is looking in the direction of the surface and it is measured in candelas per square meter (or per square foot).

Lux (lx): A unit of illuminance or light falling onto a surface. One lux is equal to one lumen per square meter. Ten lux approximately equals one footcandle.

Mercury Lamp: A high-intensity discharge light source operating at a relatively high pressure (about 1 atmosphere) and temperature in which most of the light is produced by radiation from excited mercury vapor. Phosphor coatings on some lamp types add additional light and improve color rendering.

Metal Halide Lamp: A high-intensity discharge light source in which the light is produced by the radiation from halides of metals such as sodium, scandium, indium, and dysprosium. Some lamp types may also utilize phosphor coatings.

Mesopic: Referring to nighttime outdoor lighting conditions, the region between photopic and scotopic vision.

Monochromatic Light: Light with only one wavelength (i.e., color) present.

Nominal Watts: The power rating of lamps, as published by lamp manufacturers.

Opal Glass: Milky, translucent glass produced by adding ingredients to clear glass. Used for diffusing light.

PAR Lamp: PAR is an acronym for parabolic aluminized reflector. A PAR lamp, which may utilize either an incandescent filament, a halogen filament tube, or a HID arc tube, is a precision pressed-glass reflector lamp. PAR lamps rely on both the internal reflector and prisms in the lens for control of the light beam.

Phosphor: An inorganic chemical compound processed into a powder and deposited on the inner glass surface of fluorescent tubes and some mercury and metal halide lamp bulbs. Phosphors are designed to absorb short-wavelength ultraviolet radiation and to transform and emit it as visible light.

Photopic: The vision for which the cones in the eye are responsible; typically at high brightness and in the foveal or central region.

Preheat Lamp: A fluorescent lamp in which the filament must be heated by use of a starter before the lamp starts. These lamps are typically operated with electromagnetic ballasts.

Pulse Start: An HID ballast with a high-voltage ignitor to start the lamp.

Quartz Lamp: The term derives from the quartz glass that encloses the filament and halogen gas in halogen lamps. Quartz glass can withstand the high pressure of the halogen lamp, but it transmits more UV radiation than ordinary hard glass. Touching the quartz glass with bare hands leaves an oily residue that greatly reduces lamp life.

Rapid Start Lamp: A fluorescent lamp with two pins at each end connected to the filament. The filaments are heated by the ballast to aid in starting. Some rapid start lamps may be instant started without filament heat.

Rated Lamp Life: For most lamp types, rated lamp life is the length of time of a statistically large sample between first use and the point when 50% of the lamps have died.

Reflector Lamps: A light source with a built in reflecting surface, usually a silver or aluminum coating on the bulb.

Restart or Restrike Time: The amount of time after an interruption to the point of lamp ignition.

Retina: A light-sensitive membrane lining the posterior part of the inside of the eye.

Retrofit: A self-ballasted replacement lamp that converts a light source to either change its characteristics or reduce energy consumption.

Rods: Retinal receptors that respond to low levels of luminance but cannot distinguish hues. Not present in the center of the fovea region.

Scotopic: The vision where the rods of the retina are exclusively responsible for seeing (very low luminance conditions and more sensitive to blue emissions).

Scotopic/Photopic (S/P) Ratio: This measurement accounts for the fact that of the two light sensors in the retina, rods are more sensitive to blue light (scotopic vision) and cones to yellow light (photopic vision). The scotopic/photopic (S/P) ratio is an attempt to capture the relative strengths of these two responses. S/P is calculated as the ratio of scotopic lumens to photopic lumens for the light source. Cooler sources (higher color temperature lamps) tend to have higher values of the S/P ratio compared to warm sources.

Self-Ballasted Lamps: A fluorescent lamp with integrated ballast allowing it to be directly connected to a socket providing mains voltage.

Service Lamp Life: The service lamp life is another term and is defined as the result of the multiplication of lifetime and lumens maintenance. Often 70% service life or 80% service lifetime is used. This is the number of operating hours after which, by a combination of lamp failure and lumen reduction, the light level of an installation has dropped to 70% or 80%, compared to the initial value.

Spill Light: Light that is not aimed properly or shielded effectively can spill out into areas that do not want it: it can be directed toward drivers, pedestrians, or neighbors.

Spot Lamp: Reflector lamp with a narrow light distribution.

Task Lighting: Lighting directed to a specific work area or task such as a table for reading.

Transformer: A device used to raise (step up) or lower (step down) the electric voltage. For example, many halogen lamps require a transformer to reduce mains voltage (220 volts) to low voltage.

Triphosphor: A mixture of three phosphors to convert ultraviolet radiation to visible light in fluorescent lamps with each of the phosphors emitting light that is blue, green, or red in appearance and the combination producing white light.

Troffer: Recessed fluorescent fixture for use in a suspended ceiling; derived from the words *trough* and *coffer.*

Warm Up Time to 90%: The time it takes for a high-intensity discharge lamp to reach 90% of light output after being turned on.

Warm White: This refers to a color temperature of <3500 K, providing a yellowish-white light.

Appendix A

Parts of the Electromagnetic Spectrum

Most of the electromagnetic spectrum is invisible, and exhibits frequencies that traverse its entire breadth. Exhibiting the highest frequencies are gamma rays, X-rays, and ultraviolet light. Infrared radiation, microwaves, and radio waves occupy the lower frequencies of the spectrum. Visible light falls within a very narrow range in between.

Radio Waves	10^4–10^{-2} m/10^4–10^{10} Hz	
	Ultra-low frequency (ULF)	3–30 Hz
	Extremely low frequency (ELF)	30–300 Hz
	Voice frequencies (VF)	300 Hz–3 kHz
	Very low frequency (VLF)	3–30 kHz
	Low frequency (LF)	30–300 kHz
	Medium frequency (MF)	300 kHz–3 MHz
	High frequency (HF)	3–30 MHz
	Very high frequency (VHF)	30–300 MHz
	Ultra-high frequency (UHF)	300 MHz–3 GHz
	Super-high frequency (SHF)	3–30 GHz
	Extremely high frequency (EHF)	30–300 GHz
	Shortwave	MF, HF
	Television	VHF, UHF
	Microwaves	30 cm–1 mm/1–300 GHz
Infrared	10^{-3}–10^{-6} m/10^{11}–10^{14} Hz	
	Far	1000–30 μm
	Mid	30–3 μm
	Near	3–0.75 μm

Continued

Visible	5×10^{-7} m/2 $\times 10^{14}$ Hz	
	Red	770–622 nm
	Orange	622–597 nm
	Yellow	597–577 nm
	Green	577–492 nm
	Blue	492–455 nm
	Violet	455–390 nm
Ultraviolet	10^{-7}–10^{-8} m/10^{15}–10^{16} Hz	
	UV-A (less harmful)	400–315 nm
	UV-B (harmful, absorbed by ozone)	315–280 nm
	UV-C (more harmful, absorbed by air)	280–100 nm
	Near UV (black light)	400–300 nm
	Far UV	300–200 nm
	Vacuum UV	200–100 nm
X-rays	10^{-9}–10^{-11} m/10^{17}–10^{19} Hz	
Gamma Rays	10^{-11}–10^{-13} m/10^{19}–10^{21} Hz	

Appendix B

Comparative Values of CRI and Color Temperatures
for a Variety of Light Sourcs

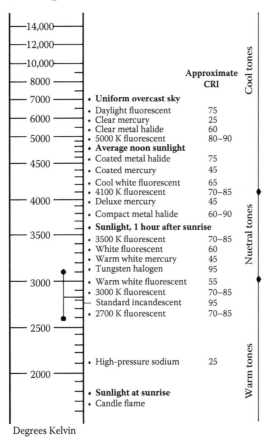

Degrees Kelvin		Approximate CRI	
14,000			Cool tones
12,000			
10,000			
8000			
7000	• Uniform overcast sky		
	• Daylight fluorescent	75	
6000	• Clear mercury	25	
	• Clear metal halide	60	
5000	• 5000 K fluorescent	80–90	
	• Average noon sunlight		
4500	• Coated metal halide	75	
	• Coated mercury	45	
	• Cool white fluorescent	65	
	• 4100 K fluorescent	70–85	
4000	• Deluxe mercury	45	Nuetral tones
	• Compact metal halide	60–90	
	• Sunlight, 1 hour after sunrise		
3500	• 3500 K fluorescent	70–85	
	• White fluorescent	60	
	• Warm white mercury	45	
	• Tungsten halogen	95	
3000	• Warm white fluorescent	55	
	• 3000 K fluorescent	70–85	
	— Standard incandescent	95	
	• 2700 K fluorescent	70–85	
2500			
	• High-pressure sodium	25	
2000			Warm tones
	• Sunlight at sunrise		
	• Candle flame		

239

Appendix C

List of Phosphor Powders for Fluorescent Lamps

$(Ba,Eu)Mg_2Al_{16}O_{27}$	Blue phosphor for trichromatic fluorescent lamps
$(Ce,Tb)MgAl_{11}O_{19}$	Green phosphor for trichromatic fluorescent lamps
$(Y,Eu)_2O_3$,	Red phosphor for trichromatic fluorescent lamps
$(Sr,Eu,Ba,Ca)_5(PO_4)_3Cl$	Blue phosphor for trichromatic fluorescent lamps
$(La,Ce,Tb)PO_4$	Green phosphor for trichromatic fluorescent lamps
$Y_2O_3{:}Eu$	Red phosphor (611 nm) for trichromatic fluorescent lamps
$LaPO_4{:}Ce,Tb$	Green phosphor (544 nm) for trichromatic fluorescent lamps
$(Sr,Ca,Ba)_{10}(PO_4)_6Cl_2{:}Eu$	Blue phosphor (453 nm) for trichromatic fluorescent lamps
$BaMgAl_{10}O_{17}{:}Eu,Mn$	Blue-green (456/514 nm)
$(La,Ce,Tb)PO_4{:}Ce,Tb$	Green (546 nm) phosphor for trichromatic fluorescent lamps
$Zn_2SiO_4{:}Mn$	Green (528 nm)
$Zn_2SiO_4{:}Mn,Sb_2O_3$	Green (528 nm)
$Ce_{0.67}Tb_{0.33}MgAl_{11}O_{19}{:}Ce,Tb$	Green (543 nm) phosphor for trichromatic fluorescent lamps
$Y_2O_3{:}Eu(III)$	Red (611 nm) phosphor for trichromatic fluorescent lamps
$Mg_4(F)GeO_6{:}Mn$	Red (658 nm)
$Mg_4(F)(Ge,Sn)O_6{:}Mn$	Red (658 nm)
$MgWO_4$	Soft blue (473 nm), wide range, deluxe
$CaWO_4$	Blue (417 nm)
$CaWO_4{:}Pb$	Blue (433 nm), wide range
$(Ba,Ti)_2P_2O_7{:}Ti$	Blue-green (494 nm), wide range, deluxe
$Sr_2P_2O_7{:}Sn$	Blue (460 nm), wide range, deluxe
$Ca_5F(PO_4)_3{:}Sb$	Blue (482 nm), wide range
$Sr_5F(PO_4)_3{:}Sb,Mn$	Blue-green (509 nm), wide range
$BaMgAl_{10}O_{17}{:}Eu,Mn$	Blue (450 nm) phosphor for trichromatic fluorescent lamps
$BaMg_2Al_{16}O_{27}{:}Eu(II)$	Blue (452 nm)
$BaMg_2Al_{16}O_{27}{:}Eu(II),Mn(II)$	Blue (450 + 515 nm)
$Sr_5Cl(PO_4)_3{:}Eu(II)$	Blue (447 nm)

Continued

$Sr_6P_5BO_{20}$:Eu	Blue-green (480 nm)
$(Ca,Zn,Mg)_3(PO_4)_2$:Sn	Orange-pink (610 nm), wide range
$(Sr,Mg)_3(PO_4)_2$:Sn	Orange-pink white (626 nm), wide range, deluxe
$CaSiO_3$:Pb,Mn	Orange-pink (615 nm)
$Ca_5F(PO_4)_3$:Sb,Mn	Yellow
$Ca_5(F,Cl)(PO_4)_3$:Sb,Mn	Daylight
$(Ca,Sr,Ba)_3(PO_4)_2Cl_2$:Eu	Blue (452 nm)
$3\ Sr_3(PO_4)_2.SrF_2$:Sb,Mn	Blue (502 nm)
$Y(P,V)O_4$:Eu	Orange-red (619 nm)
$(Zn,Sr)_3(PO_4)_2$:Mn	Orange-red (625 nm)
Y_2O_2S:Eu	Red (626 nm)
$(Sr,Mg)_3(PO_4)_2$:Sn(II)	Orange-red (630 nm)
$3.5\ MgO.\ 0.5\ MgF_2.\ GeO_2$:Mn	Red (655 nm)
$Mg_5As_2O_{11}$:Mn	Red (660 nm)
$Ca_3(PO_4)_2.CaF_2$:Ce,Mn	Yellow (568 nm)
$SrAl_2O_7$:Pb	Ultraviolet (313 nm)
$BaSi_2O_5$:Pb	Ultraviolet (355 nm)
$SrFB_2O_3$:Eu(II)	Ultraviolet (366 nm)
SrB_4O_7:Eu	Ultraviolet (368 nm)
$MgGa_2O_4$:Mn(II)	Blue-green, black light
$(Ce,Tb)MgAl_{11}O_{19}$	Green

Appendix D

Aquariums

- Neutral and cool white metal halides
- Neutral and cool white fluorescent
- Neutral and cool white LEDs

Art studios

- Halogen lamps
- Fluorescent lamps with high CRI
- LED lamps with high CRI

Astronomical institutions

- Sodium lamps for nearby outdoor areas

Barcodes

- Mercury discharge quartz lamps
- Fluorescent UV lamps
- LED UV lamps

Bathrooms

- Halogen lamps
- Fluorescent lamps
- LED lamps

Bedrooms

- Halogen lamps
- Warm white fluorescent
- Warm white LED

Blood treatment

- Red LED

Bridges

- Cool white metal halide lamps

- Cool white fluorescent lamps (inductive lamps offer longer lifetimes)

- Cool white LED systems

Bug zappers

- Fluorescent UV lamps

- LED UV lamps

Cataracts

- Halogen lamps

- Warm white fluorescent lamps (high CRI)

- Warm white LED lamps (high CRI)

- Use UV filters and minimize blue emissions

Cell imaging

- Fluorescent UV lamps

- LED UV lamps

Churches

- Halogen lamps

- Warm white fluorescent

- Warm white LED

Closed circuit television (CCTV)

- Halogen lamps

Color vision impairment

- LED lamps of appropriate colors

- Fluorescent lamps of appropriate colors

Concert halls

- Halogen lamps
- Warm white fluorescent
- Warm white LED

Corridors

- LEDs

Curing of polymers and printer inks

- Fluorescent UV lamps
- LED UV lamps

Darkrooms (photography)

- Red LEDs

Decoration

- LEDs (white or specific colors)

Delayed sleep phase syndrome

- Cool white fluorescent and LED lamps
- Blue and blue-green fluorescent and LED lamps

Dentistry

- Halogen lamps
- Fluorescent (high CRI)
- LED lamps (high CRI)

Disinfection

- Mercury discharge quartz lamps

DNA sequencing

- Fluorescent UV lamps

Drawing offices

- Halogen lamps

- Fluorescent (high CRI)

- LED lamps (high CRI)

Drug detection

- Fluorescent UV lamps

Drug user areas

- Blue LED lamps

- Blue fluorescent lamps

Dry heating

- Halogen lamps

Dusty areas

- Sodium lamps

- Dynamic lighting

- Fluorescent lamps (warm and cool white)

- LEDs lamps (warm and cool white)

- Colored fluorescent or LEDs for a variety of ambient lighting

Eczema

- Fluorescent UV lamps

Emergency lights

- Colored LEDs

Fabric industry

- Halogen lamps

- Fluorescent lamps with high CRI

- LED lamps with high CRI

Flash

- White LEDs
- Low-pressure xenon lamps

Flicker

- Warm white fluorescent lamps with electronic gear
- Pulsing warm white LED systems

Florists

- Halogen lamps
- Fluorescent lamps with high CRI
- LED lamps with high CRI

Foggy areas

- Sodium lamps

Food heating areas

- Halogen lamps

Food processing

- Halogen lamps
- Coated fluorescent (high CRI)

Footpaths

- LEDs—solar lamps

Forensic analysis

- Fluorescent UV lamps

Galleries

- Halogen lamps

- Fluorescent lamps with high CRI

- LED lamps with high CRI

Gardens

- LEDs—solar lamps

Glaucoma

- Halogen lamps

- Warm white and neutral white fluorescent lamps (high CRI)

- Warm white and neutral white LED lamps (high CRI)

- Use UV filters and minimize blue emissions

Gym and exercise rooms

- Cool white fluorescent lamps

- Cool white LED lamps

Hair growth

- Red LED

Hair removal

- Xenon flash lamps

Heating units

- Halogen lamps

Heavy industry

- Sodium lamps

Horticulture

- Neutral and cool white metal halides

- Neutral and cool white fluorescent

- Neutral and cool white LEDs

Hospitals

- Halogen

- Fluorescent (high CRI)

- LED (high CRI)

Hotels

- Halogen

- Warm white fluorescent

- Warm white LED

Industrial manufacturing (high temperatures)

- Halogen lamps

Industry

- Metal halide lamps

- Fluorescent (inductive for longer lifetimes)

Kitchens

- Halogen

- Coated fluorescent (high CRI)

Label tracking

- Mercury discharge quartz lamps

- Fluorescent UV lamps

Living rooms

- Halogen lamps

- Warm white fluorescent

- Warm white LED

- LED systems for dynamic lighting

Make-up rooms

- Halogen lamps

- Fluorescent lamps with high CRI

- LED lamps with high CRI

Melatonin suppression

- Cool white fluorescent and LED lamps

- Blue and blue-green fluorescent and LED lamps

Motion detection systems (instant starting)

- Halogen

- LEDs

Museums

- Halogen lamps

- Fluorescent lamps with high CRI

- LED lamps with high CRI

Neonatal jaundice

- Blue and blue-green fluorescent

- Blue and blue-green LED lamps

Night home lighting

- Halogen lamps

- Warm white fluorescent and LED lamps

- Red light LED

Offices

- Fluorescent lamps

- LED lamps

- Color temperature depends on task

Open roads

- Sodium lamps

Pain management

- Red LED

Parking lots

- Cool white metal halide lamps
- Cool white fluorescent lamps (inductive lamps offer longer lifetimes)
- Cool white LED systems (long lifetimes)

Photophobia

- Dimming halogen or LED systems

Poultry brooding

- Halogen lamps

Printing facilities

- Halogen lamps
- Fluorescent lamps with high CRI
- LED lamps with high CRI

Protein analysis

- Fluorescent UV lamps

Psoriasis

- Fluorescent UV lamps

Reception areas

- Halogen lamps
- Warm white fluorescent
- Warm white LED

Reptile housings

- Halogen lamps

- Fluorescent UV lamps

Restaurants

- Halogen lamps

- Warm white LED

- LED systems for dynamic and varying atmospheres

Retinitis pigmentosa (RP)

- Halogen lamps

- Warm white and neutral white fluorescent lamps (high CRI)

- Warm white and neutral white LED lamps (high CRI)

- Use UV filters and minimize blue emissions

Saunas

- Halogen lamps

Schools

- Neutral white fluorescent (inductive for longer lifetimes)

- Neutral white metal halide lamps

Seasonal affective disorder (SAD)

- Cool white fluorescent lamps

- Cool white LED lamps

- Blue and blue-green fluorescent and LED lamps

Senior citizen homes

- Halogen lamps

- Warm white fluorescent

- Warm white LED

Ships

- LEDs

Shops

- Halogen lamps

- Fluorescent lamps

- LEDs

Signs

- LEDs

Sinus-related diseases

- Red LED

Skin marks removal

- Xenon flash lamps

Skin tanning

- Fluorescent UV lamps

Squares (green spaces)

- Cool white metal halide lamps

- Cool white fluorescent lamps (inductive lamps offer longer lifetimes)

- Cool white LED systems (long lifetimes)

Stadiums

- Cool white metal halide lamps

- High-pressure xenon

Stages

- Halogen (and filters)

- High-pressure sodium

- High-pressure xenon

- LEDs

- Metal halide (high CRI) lamps

Stairs

- LEDs

Sterilizing

- Mercury discharge quartz lamps

Storage rooms

- LEDs

Streets and roads

- Cool white metal halide lamps

- Cool white fluorescent lamps (inductive lamps offer longer lifetimes)

- Cool white LED systems (long lifetimes)

Studios

- Halogen (and filters)

- High-pressure sodium

- High-pressure xenon

- LEDs

- Metal halide (high CRI) lamps

Study rooms

- Cool white fluorescent

- Cool white LED

Theaters

- Halogen lamps

- Warm white fluorescent

- Warm white LED

Tunnels

- Sodium lamps

Ultraviolet sources

- Halogen quartz lamps

- Mercury discharge quartz lamps

- Fluorescent UV lamps

UV protection

- UV filters in all halogen or discharge lamps

- Non-UV LED lamps

UV-ID

- Mercury discharge quartz lamps

Vitiligo

- Fluorescent UV lamps

Waiting rooms

- Halogen lamps

- Warm white fluorescent

- Warm white LED

Warehouses

- Sodium lamps

- Metal halide lamps

Water treatment

- Mercury discharge quartz lamps

Worship places

- Halogen lamps

- Warm white fluorescent

- Warm white LED

Wound healing

- Red LED

Wrinkle removal (photo-rejuvenation)

- Xenon flash lamps

Index